PENGUIN CLASSICS

AUTOBIOGRAPHIES

CHARLES DARWIN was born into an upper-middle-class medical family in 1809. He was destined for a career in either medicine or the Anglican Church. However, he did not complete his Edinburgh medical education and after leaving Christ's College, Cambridge, in 1831, his future was entirely changed when he joined HMS *Beagle* as a self-financing, independent naturalist. On returning to England in 1836 he began to write up his theories and observations which culminated in a series of books, most famously *On the Origin of Species by Means of Natural Selection* of 1859. He married his cousin Emma Wedgwood in 1839. In 1842 they moved to Down House in the north Kent countryside where Darwin lived for the rest of his life. During this time he was socially reclusive and continually ill. He died in 1882 and was buried in Westminster Abbey.

MICHAEL NEVE was born in Tokyo in 1949. He graduated in history from Christ's College, Cambridge, in 1971 and is currently based at the Wellcome Trust Centre for the History of Medicine at University College London. He teaches and researches the history of psychiatry and the history of the life sciences. With Janet Browne he co-edited Darwin's *Voyage of the Beagle*, also for Penguin Classics.

SHARON MESSENGER was born in Cirencester in 1971. She graduated in social history from the University of Liverpool in 1992 and completed her Ph.D. in 1999. She is a research officer at the Wellcome Trust Centre for the History of Medicine at University College London.

T0200934

CHARLES DARWIN

Autobiographies

Edited by MICHAEL NEVE *and* SHARON MESSENGER
With an Introduction by MICHAEL NEVE

PENGUIN BOOKS

PENGUIN BOOKS

Published by the Penguin Group
Penguin Books Ltd, 80 Strand, London WC2R ORL, England
Penguin Putnam Inc., 375 Hudson Street, New York, New York 10014, USA
Penguin Books Australia Ltd, 250 Camberwell Road, Camberwell, Victoria 3124, Australia
Penguin Books Canada Ltd, 10 Alcorn Avenue, Toronto, Ontario, Canada M4V 3B2
Penguin Books India (P) Ltd, 11, Community Centre, Panchsheel Park, New Delhi – 110 017, India
Penguin Books (NZ) Ltd, Cnr Rosedale and Airborne Roads, Albany, Auckland, New Zealand
Penguin Books (South Africa) (Pty) Ltd, 24 Sturdee Avenue, Rosebank 2196, South Africa

Penguin Books Ltd, Registered Offices: 80 Strand, London WC2R ORL, England

www.penguin.com

'An autobiographical fragment' first published in 1903, and 'Recollections' first published in 1887
First published in Penguin Classics 2002

027

'An autobiographical fragment' copyright © Cambridge University Press, 1986
'1876 May 31 – Recollections of the Development of my Mind and Character' copyright © George Pember Darwin, 2002
Editorial material copyright © Michael Neve and Sharon Messenger, 2002

Set in 10.25/13 pt Monotype Bulmer
Typeset by Rowland Phototypesetting Ltd, Bury St Edmunds, Suffolk
Printed and bound in Great Britain by Clays Ltd, Elcograf S.p.A.

ISBN 978-0-14-043390-6

www.greenpenguin.co.uk

CONTENTS

ACKNOWLEDGEMENTS

I should like to thank both colleagues and students at the Wellcome Trust Centre for the History of Medicine at University College London, with special thanks to Dr Janet Browne. Darwin is fortunate to have her as a biographer. Paul Keegan, late of Penguin Classics, set up the project and Robert Mighall and then Laura Barber at Penguin kept me at it. John Sturrock exchanged his ideas on the language of autobiography with me when he was writing on that subject some years ago and I gained a great deal from him. Thanks to all. Caroline Essex was a meticulous proof-reader in the final stages of this edition. Dr Sharon Messenger, also based at the Wellcome Trust Centre at UCL, started out doing research assistance but had her own perceptions of Darwin's memoirs and ended up a most agreeable combination of researcher and co-editor. For revealing – and often very funny – conversations about telling stories of one's own life, and what a father in particular might best leave in and best leave out, special thanks to Flora and Georgia Neve and to Sarah and Grace Benton.

Grateful acknowledgements are made to the Cambridge University Press for permission to use the printed 'Fragment', to the Syndics of Cambridge University Library for access to and the reprinting of DAR 91: 56–62, and to George Pember Darwin, and for the 'Recollections' to the Syndics of Cambridge University Library and again to George Pember Darwin.

CHRONOLOGY

1809	12 February: Born in Shrewsbury, the son of the physician Robert Waring Darwin and Susannah, née Wedgwood
1817	Starts at a day school run by Mr G. Case, Unitarian minister
1817	July. Darwin's mother dies
1818	Becomes a boarder at Shrewsbury School
1825-7	Medical studies at Edinburgh University; leaves without completing his degree
1828-31	January 1828: Goes up to Christ's College, Cambridge. Passes BA without Honours (1831)
1831-6	HMS *Beagle* departs from Plymouth; after circumnavigating the globe the *Beagle* returns to Falmouth
1839	29 January: Marries his cousin Emma Wedgwood (1808-96); they had ten children
1842	Settles with his wife and family in Down, Kent
1850s	Works on a long study of natural selection and related topics. Only the realization that others were working in similar areas drives Darwin to publish the *Origin of Species* in November 1859
1871	*The Descent of Man and Selection in Relation to Sex*
1872	*The Expression of the Emotions in Man and Animals*
1881	*The Formation of Vegetable Mould through the Action of Worms*, Darwin's last book
1882	19 April: Dies; concerted efforts by Fellows of the Royal Society lead to his being buried in Westminster Abbey

INTRODUCTION

In the year 1879, having already begun to make additions to the autobiographical recollections he had started in the spring and summer of 1876, Charles Darwin wrote to a fellow scientist whom he trusted. Trusted with that most delicate of tasks – translation. Julius Victor Carus was a zoology professor at Leipzig and a translator of both Darwin and Darwin's friend and ally T. H. Huxley. The delicacy of the business of translation mattered acutely to Darwin, because he did not wish such translations to risk controversy, or to exaggerate the materialist and radical tone of his works and thereby encourage social and political debate. Carus could manage that neutrality of tone. But Darwin made one thing clear to his German colleague. He would never dream of publishing his autobiography.

Darwin was indeed working on an account of himself, or rather himself as embodied in his writings, but it absolutely was not meant for the public eye. Famously connected to the scientific account of evolution by means of natural selection, and an evangelical proponent of its truth, Darwin in the twenty-first century is less well known for a range of other studies in natural history and for an enormous personal correspondence. Less well known, one might say, as a writer of more than the one book, the *Origin of Species* of 1859, that made his name. But as the author of an autobiography, Darwin would have been displeased that his private reflections could be read or that various editions of them now exist. Although the first version of the autobiography and then the first complete one were edited by his immediate family or descendants, we shall never know for certain what he would have thought of this familial decision to let the cat out of the bag. But almost certainly he would have taken the view that a private world was being entered while the body of his life's work and its public published status was ignored or under-studied. The depth of the implications of the theory of evolution by natural selection was being exchanged – for gossip.

Darwin died before Queen Victoria, under whose reign almost his entire scientific career had taken place. Not only do his views on decorum and good conduct have 'Victorian' aspects, but he also combined his personal wealth and the security it brought him with a very Victorian combination of privacy and illness and endless hard work. Darwin worked: indeed he almost did nothing but. When not working, he was ill. But this powerful blend of labour and illness took place behind closed doors. He was intensely ambitious for his theory and a surprise letter coming through the door, indicating similar ambitions on the part of other naturalists, would spur him into immediate action to establish his priority. But, equally, had he not heard from these other writers and naturalists, he might have remained silent on his own researches and his own conclusions for decades or even longer. His wealth helped him avoid deadlines if he so chose but it also enabled him to have the contacts and the support never to be late for a deadline he had agreed. It was a matter of balancing the books, of combining rest with anxious labour, of pros and cons, pluses and minuses. Darwin was a very careful balancer of his private financial books, and this sense of balance and the weighing up of the good against the bad is evident in his autobiographical reflections.

The public availability of this private text would certainly have been a minus. And who could be trusted to edit such a document? True, it was another German editor, one Ernest Von Hesse Wartegg, then exiled in Paris, who helped begin the act of recollecting by writing to the great man in 1875, asking for some biographical information. But Darwin meant the main text reprinted here to be a private document, where he gave an account of his own character and development that his family might read and reflect on and in which they might see something of themselves – written as it was by the patriarchal head of the family. The longer of these two pieces was meant for his children and grandchildren at a time when he was full of concerns. He worried that his children as a group were so late in producing offspring. He worried about their health because his own health was dreadful and he might have passed that on to them. Darwin had committed himself to an overwhelmingly hereditarian view of the natural order and of human nature. But might there be a dreadful sting in the tail – that he himself was a suspect and not a healthy and responsible ancestor? Was he an agent of bad heredity? For a man who had argued that Malthusian forces made natural selection the engine of all history, that Malthus on population had set out the true model whereby population always outran resources and where selection had to eliminate the many and leave enough for the few, there could be no more important concern. Were the Darwins to be among the favoured victors in life's battles? And might he as the father set out a story of how he personally had achieved what he had and thereby assist his children and their children to do likewise?

Darwin had indeed become alarmingly famous, as a result of his authorship of books and his study of the natural world. But how had that happened and what lessons – lessons about work and patience and no distractions – had been learned? The shorter autobiographical piece, which appears first in this collection, is a fragment from the summer of Darwin's twenty-ninth year and has no didactic purpose comparable to his recollections of the late 1870s. Instead the 'fragment', as it was called, was penned at a time when Darwin was pondering the implications of a growing evolutionary perspective on human origins and human psychology. In his 'M and N notebooks' of the late 1830s he began to mull over this and in doing so began to consider his first memories of anything at all. Why *those* memories? And what was selecting them? In the second, longer and more explicitly autobiographical document, there is a private and a familial purpose at work and it is impossible to interpret the autobiography historically without always recalling the context and the intended audience. But add to that the size of the current Darwin industry, its academic armies swarming over every tiny detail of Darwin's life, and the author's likely reaction is even clearer. Darwin's short autobiographical writings have already received a historical going-over that would have surprised and infuriated him. Attention to his scientific publications was one thing, but a family memoir?

Darwin's strategy as an author writing in the first-person singular is to persuade his family and especially his male descendants that there are no real lessons in how to succeed except hard work, observation and – in his case – the luck, social contacts and private money to undergo one great and determining event – the journey on HMS *Beagle*. As the best recent scholars of the story have shown, the adult Darwin was a valetudinarian and a recluse, finding a world in a path of sand at his Kent home, rarely straying except for holidays and odd visits to London and properly departing only when ill and needing medical treatment, probably at a spa. Much of the tone and the various humilities in the text – humilities not all to be believed – are present as part of Darwin's need to set a proper example for his immediate family. Here we have diffidence, surprise at world fame, a self-imaging around the claim that good observation, patience and a certain ordinariness of mind produce true 'scientific' results. No genius he, according to him. Instead he insisted on the wisdom of 'It's dogged as does it'. Indeed, Darwin makes claims for the ordinariness of proper scientific labour and for a form of resilient, unglamorous daily labour as the foundation stone for true science. The inspired nature of genius would be fine while it lasted but would be bound to fade – a Romantic shooting star. The real backbone for something to be scientific, as against fantastical or speculative or just plain wrong, had to be daily, lacklustre labour. To stray from that path, as for example the evolutionist Alfred Russel Wallace did when taking up spiritualism in the 1860s, was to court disaster.

Darwin does not, however, make all that much in these pages of the prolonged and still mysterious illness that determined much of his routine and his daily experience for over four decades. But if we remember the context – writing for the familial reproductive community with the father both protecting his reputation and setting an example – then the awkwardness and the exaggerated diffidence of some of his sentences make sense. He is, in his own eyes, setting an example and any suggestion that his work had made him chronically ill would have been inappropriate. There is almost an ideology of honesty and modesty being deployed here, because of the family audience. And that family focus is exemplified by the starting of his recollections in his cousin's house. Cousins and the marrying of cousins (how safe was this from a hereditarian point of view?) was a serious concern of Darwin's – he had, after all, married his cousin Emma Wedgwood. The subject also became one of the foci of the later academic work of his son George. It seems entirely in keeping that most of these pages were started at Hopedene, home of Darwin's cousin Hensleigh Wedgwood and his wife Fanny. At the heart of everything, everything that mattered, everything that was real, lay family life, family structure, familial inheritance, familial role models. A particularly Victorian claustrophobia is evident here.

The family's response to these pages was not straightforward. Acting as the first of the editors (and censors), Darwin's wife Emma and especially their son Francis (Frank) found some of what was written distressing: opinions on character and the harsh views expressed on Christianity. They agreed to remove these parts before publication and Francis Darwin prepared an expurgated version for publication in 1887. Once again it was a matter of balance. The world needed to have the memory of the man kept alive but in a carefully selected way. Brought into the public world in a familially censored form, what does the writing presented here allow us to see? Since we have the text and have given in to the temptation to read it and judge it, what do we find? How does it hide while we seek? One important and telling event appears in both the fragment and the longer work of Darwin's later life – the death of his mother. This event was sufficiently striking for the eminent British child psychiatrist John Bowlby to write a thorough biography of Darwin in which the loss of the mother, who died when Darwin was eight, determines the illnesses, anxieties and longings for domestic peace of Darwin's adult life. Equally his marriage to his cousin has been construed as a maternal rather than a passionate or erotic relationship. In these pages, Darwin's mother does not have a corporeal existence. Rather like some of the illustrations made of Darwin's study at Down House at the height of his fame, where the sage is always absent from the book- and instrument-filled room, the memories related to the mother are of things and not persons: the room where she died, trivial conversations, her gown, her work-table and death-bed. Darwin lived in an age before the human and medical sciences had developed the modern rhetoric of 'repressed longings', and before Freud had defined the Oedipus complex. Nothing divides Darwin's text from modernity as much as the simple claim that he did not know his mother well, did not miss her and did not seek her in later female loves. But for modern psychiatry as well as psychoanalytic theory, this ghost casts a long shadow. This maternal absence is a very real presence. Depending on the individual reader's sense of both psychology and human character, Darwin's account of his mother speaks of acute, albeit hidden longings and a lifelong search for maternal substitutes and a safe, secluded life. Bowlby proposes that Darwin's vulnerable personality was the result of a childhood shadowed by an invalid mother and an often intimidating father, and that the range of symptoms in his illness – vomiting, exhaustion and eczema among them – were responses to stress. Bowlby also suggests that familial silence on the mother's death made Darwin highly sensitive to any illness or death in his own family. It also meant that he hid strong feelings – anger or pride – and developed hyperventilation syndrome and depression, the sick roles for the motherless child.

For those who take Darwin at his word, who look no further and trust his truthfulness, his memories speak of realism and a complete absence of sentimentality. Or – to take the title of a poem by William Wordsworth, whom Darwin liked but along with Shakespeare could not read in later life – of resolution and independence. A psychiatrist such as Bowlby has a very particular axe to grind on the consequences of maternal deprivation. That Darwin was essentially truthful on the matter of his mother's death and that there is no further analysis to be made can seem to the modern interpreter naïve and evasive. Darwin was – in the current and ugly usage – 'in denial'. But seeing him as 'in denial' does not help explain the times when Darwin is overt and detailed. On his father's anger Darwin is explicit. He writes here that his father was 'easily made very angry'. Darwin even suggests that Dr Robert Darwin made use of this capacity, this sense of being about to be angry, in his medical practice and in the doctor/patient relationship. In ways that became famous in the case of the madness of King George III, doctors treating the mentally disturbed in the late eighteenth century made use of what were called 'moral' therapies. Instead of physically restraining the insane, they resorted to non-physical or 'moral' means: fear, surveillance, rewards and punishments, even staring into the patient's eyes. The patient was in the grip of the doctor's theatrical power, by turns frightening or kind or mysterious. Dr Darwin is portrayed by his son as using fear as part of a patient's therapy. Explicitly, Dr Darwin used the threat of turning up on the doorstep if the insane wife of one 'Major B——' stepped out of line. No reader can fail to be struck by how straightforward Darwin is about his father's anger, his insistence that he got his own way (Charles was forced to read medicine until he could take no more and dropped it, an act requiring some courage) and also quotes a sentence of paternal dismissal and rage that bears repetition. Fearing his son was already on the road to nowhere in his early youth, Dr Darwin said to him: 'You care for nothing but shooting, dogs, and rat-catching, and you will be a disgrace to yourself and all your family.' By the time Charles himself was a father, he can deem this appropriate where once he must have found it excessive. He writes in the autobiography that his father was 'very properly vehement against my turning an idle sporting man' and seeks to teach a lesson for the future by agreeing with this judgement from the past. Father was right then and will be now. But it would be very odd, even in an age of psychiatric explanations for almost all human action (and inaction), not to accord Darwin the credit for telling it like it is. His father was a bully and that needed saying. And there was no mother for him to remember. There was another reason for his silence (a silence very much not the same thing as repressing the memory of loss) – respect for his sisters and how they grieved. This remembrance of things (but not persons) past was also at work at the burial of a soldier he attended. Darwin misremembered him as a dragoon soldier – he was in fact a hussar – but what he recalls perfectly clearly were the empty boots and carbine suspended from the saddle.

The unknown mother. The unknown soldier. But the longest section (written separately in 1878) on the known father did not only describe the father's anger. Darwin also balances anger against kindness, writes of the father who melted, who conceded – who tolerated Darwin's defection from learning medicine in Edinburgh and eventually agreed that the journey on the *Beagle* was as Josiah Wedgwood deemed it: a rare opportunity. The *Beagle* journey was indeed, as Charles writes here, 'the most important event' in his life and determined his 'whole career'. From the *Beagle* journey of 1831-6 onwards, Darwin insisted he was self-taught, persistent and persevering. If the financial safety that his father had secured for him meant he could both travel as a gentleman companion on the *Beagle* and then take decades to ponder the legacy of that voyage, so be it. If family money allowed him to hide from the world after he had gone around the world, then he must take the opportunity that money brought: to be both inside the world of science and, when it suited him, outside it.

Darwin's account of the voyage on the *Beagle* as it appears here is radically and suspiciously individualist. He makes no mention of collective work or the Admiralty-based network that gave safe and rapid passage to his letters and the natural history items he sent home. It has taken the historians of nineteenth-century science to restore the full social and imperial context for the *Beagle* journey and one cannot but be struck by how solitary he wished to appear for his family in reworking the *Beagle* saga for the autobiography. There is one strange presence, that of the captain of the vessel, Robert FitzRoy. Darwin's version of the journey in the autobiography unites the momentous with the silly and the accidental, combines noble labours with petty quarrels. It also gives evidence of a stubborn and individualist Darwin, of a man prepared to eliminate the history of collective effort from his autobiography and instead to paint a picture of personal effort and personal triumph. The silliness – although we laugh at our peril, having modern equivalents all around us – was that after his first meeting with FitzRoy in London, Darwin was very nearly unacceptable to the captain because of his nose. FitzRoy's judgement of Darwin's nose was based on the physiognomical system of the eighteenth-century Swiss clergyman, J. C. Lavater, who taught how character might be deduced from the face. On Lavaterian grounds, Darwin's nose was the wrong shape and suggested that he did not have the strength and determination to last the journey. FitzRoy came close to rejecting this self-financing, upper-middle-class gentleman companion, nasally challenged as he was. But he needed the company and Darwin travelled as an independent naturalist, able to leave and rejoin the vessel when he wished.

The momentousness of the *Beagle* journey in the history of evolutionary science does not need stressing. Yet missing here is the character of FitzRoy himself and something of the intensity of the relationship between the two men. FitzRoy's uncle was the one-time secretary of state for war and later foreign secretary, Robert Stewart, Viscount Castlereagh. A complicated and driven politician, Castlereagh when angered was driven to the edge (he duelled with another foreign secretary, Canning, in 1809, when he felt deceived by him); he was also the embodiment of the war against Napoleon and the sly and tyrannous organizer of a domestic war against alleged political subversives in the late eighteen-teens. (It is Castlereagh's mask that murder wears in Percy Bysshe Shelley's poem *The Mask of Anarchy* of 1819.) The captain of the vessel upon which Darwin sailed was a relative who shared some of the characteristics for which Castlereagh was famous – he was also a hard-working, competitive, reactionary Tory aristocrat with a short fuse and a melancholy heart. The short fuse was something Darwin witnessed early on on the *Beagle* – not least in FitzRoy's firm belief in the need to flog sailors. Castlereagh committed suicide in 1822, using a penknife to cut his throat. He was deemed to have done so while unsound of mind. Much later, in April 1865, Robert FitzRoy, also exhausted by work and by feelings of being marginalized and underrecognized, killed himself, cutting his throat with a razor.

Writing the autobiography in the full knowledge of all this, Darwin does not hold back. Whether it was FitzRoy defending slavery, which Darwin hated all his life, or finding fault with the crew on almost a daily basis, or abusing the absent Darwin behind his back – in six short pages we have both an account of and a warning about a foul temper bordering on pathology, mixed with a recognition that the captain also had nobility and generosity of spirit. Given the intensity of Darwin's feelings about slavery and its evils, the portrait of FitzRoy here is the closest to a political reading of character that Darwin deems fit to pass on to his family. FitzRoy's hostility to the argument of the *Origin* comes over as conventional and a response that Darwin was used to.

No mention is made here of FitzRoy's own account of the *Beagle* voyage, and over the following years Darwin increasingly began to see his own contribution as quasi-independent. Having originally been printed in 1839 as the third volume of FitzRoy's account of the voyage, Darwin's section, originally called 'Journal and Remarks 1832–36', then became 'Journal of Researches' and eventually what we now call the *Voyage of the Beagle*. But in the years between 1837 and 1839 and then with the appearance of a preface by Darwin in the 1839 'Journal of Researches', that gave no full acknowledgement to the help of others, the exchanges between FitzRoy and Darwin were fierce. FitzRoy wrote in November 1837 that Darwin was effectively eliminating the names of officers who had held the ladder that allowed him to ascend to fame and who had made no claim on any natural history specimens collected but who had given Darwin preference. FitzRoy received hardly any thanks, neither for taking Darwin on board nor for supporting him for five years. Darwin added a later sentence or two to calm things down but the determination to cast himself as solitary and self-creating persisted over the next four decades. No one reading the autobiography would know anything of FitzRoy's views on geology or natural history or meteorology. Capable of sentimentalizing other scientists in the autobiography, most notably the Cambridge botanist and clergyman J. S. Henslow, Darwin was obviously able neither to forgive nor forget in the case of FitzRoy and we can only assume the same was true for the *Beagle*'s captain.

This combination of believing in his own certainty and its simplicity produces similar myth-making in Darwin's account of science. Much of this present collection describes the genesis of his published works; he has the volumes to prove the outcome of his labours. On the matter of what truly constituted 'science' and then the reception of his work in the scientific community, Darwin was straightforward and deliberately naïve. He claimed that reviewers of his works were honest even when he was grossly misrepresented, bitterly opposed or ridiculed. They acted in good faith. Exceptions were made when the reviewer was 'without scientific knowledge', in which case the review was 'not worthy of notice'. There is a special and complicated case of anger against the Catholic anti-evolutionist St George Mivart, but the definition of proper 'science' and 'scientists' in these pages is a carefully worked out combination of a) those who elicit fair treatment even when deluded; b) those not worth thinking about; and c) (that modesty problem again) the proviso that even when received well, Darwin was 'over and over again greatly overpraised'.

Darwin the recluse is having it all ways in these short paragraphs. He is a scientist but eschews praise. He knows why others are not scientists, and asserts they are not, without spelling out why. (There was no signed copy of the *Origin* for his old Edinburgh teacher Robert Grant. Grant was a Lamarckian and became a metropolitan professorial subversive at University College London. He was no longer the acceptable face of 'science' and indeed Darwin sees him as having ceased doing 'science', inexplicably, when in London.) And there are times when the reader might conclude something rather shocking, indeed embarrassing, given that this is the famous Charles Darwin speaking on the core topic of science: it is science when he says it is and something else again when he says it isn't. And that's that. The unquestioned assumption of the scientific truth of his own work is striking, and in a family memoir sets exacting standards, the point being that the development of his own recollected self culminated in proper science and provided the backbone, the continuity, the integrated self. True science costs not less than everything. What it leaves is autobiography.

Darwin characterized some of the famous men of the day, almost always from personal experience. Here – and especially in the case of the botanist Henslow – a version of pastoral is sometimes the result, with admiration for the older generation of clergyman-naturalists and more judgemental accounts of modern snobberies, social climbing and outright professional hostility among the practitioners of metropolitan science. As to the exponents of Grand Theory – the self-important sociologist Herbert Spencer ('I think that he was extremely egotistical') and the sneering Scottish essayist Thomas Carlyle ('I never met a man with a mind so ill adapted for scientific research') – they get short shrift. Darwin was not simply working at home and avoiding the social scene and these dubious denizens inhabiting it because of his chronic ill-health. He always saw himself as self-taught, with his proper opportunities for research and writing having roots outside educational institutions and their formal curricula. Surely he was part of the social order that supported public schools, Oxbridge, the Royal Society and so on. Yet he claims he learnt nothing at Shrewsbury School, nothing in the formal lectures at Edinburgh and precious little at Christ's College, Cambridge. The Cambridge judgement is especially interesting. For Darwin it was pretty clear that the orthodox education to be had there in the late 1820s was not up to much. Christ's College was a non-event, whereas in later years a reformed and more scientific Cambridge University became almost the definition of modern learning. In his youth, neither university nor public school, in their formal curricula, provided anything worthwhile. Auto-didacticism was evident throughout Darwin's life and he felt no disloyalty in putting its merits on record. What had saved him were teachers operating outside the educational frame and outside the formal degree structure of medicine or the Cambridge BA: Robert Grant at Edinburgh, John Stevens Henslow and the geologist Adam Sedgwick at Cambridge. The notion that he was a very ordinary schoolboy, 'below the common standard in intellect' as he writes here, stayed with Darwin throughout his life. He built his alternative working milieu outside the established places, eventually establishing a domestic milieu where an endless, often tedious, scientific method – collecting, classifying, pondering, observing, writing, corresponding – might carry on for decades. His was not the world of inspired thoughts and flashes, but of grind and yet more grind. This was how the fortress of scientific knowledge could be built, day in day out, season by season. Darwin's scientific individualism was a combination of self-help, upper-middle-class Victorianism and a deliberately middle-brow, even low-brow 'ordinariness'. But the result would be something built to last. It would not be speculative, such as the evolutionary writings of his grandfather Erasmus Darwin, but intensely detailed and with a powerful and unavoidable motor: struggle, selection, and further struggle.

By the end of his life Darwin had paid the price for a lifetime of science, finding most reading in poetry or drama unbearable and lifeless. He simultaneously fulfilled himself scientifically and became a philistine. Indeed he often speaks of science – and its absence in the work of others

– as if science were based on the necessary absence of any other form of consciousness. Day by day, hour by hour, the building was pieced together. Even the best of his friends and teachers could be diverted from proper scientific work – by philanthropical commitments in the case of Henslow, or in the case of Huxley by time-consuming labours in literary work and 'efforts to improve the education of the country'. Darwin acknowledges the merits of philanthropic work and claims that his ill-health and the need to concentrate on his work prevented his greater involvement in philanthropy. However, in a harsher light and remembering his sneers at the social aspirations of the geologists Murchison and Lyell, Darwin's individualism is plain to see. The only fertile context for true science was one untouched by any diversion. Fellow human creatures were to be the beneficiaries of a scientific account of their origins, not the beneficiaries of frivolity or unplanned social calls, let alone parties and balls.

It is not difficult when reading Darwin's letters to imagine him receiving visitors at Down but thinking to himself all the time, 'I wish this were over and they were gone'. When early colleagues and co-workers moved away or defected – when Alfred Russel Wallace took up spiritualism, or Sir Charles Lyell backed away from the full force of the animal ancestry for mankind – Darwin was left with his true companions: his illness and his work. Only one daguerreotype and one chalk drawing among the many illustrations or photographs of him have another person present. In all the remainder Darwin is alone.

His correspondence was huge, his family loving and there was his dog Polly, who was to die within days of her master. Finally there was Emma, whom he mythologizes in a sentence almost incomprehensible to the modern ear: 'She has been my greatest blessing, and I can declare that in my whole life I have never heard her utter one word which I had rather have been unsaid.' The pain that Darwin felt through disagreements with his wife over Christianity is well known. Read in one way, that sentence both shields Emma from sight – she is Darwin's perfect wife – and also adds to the sense of his solitude. Their marriage was only partly their children's business and certainly none of ours. But we can still imagine Charles alone when Emma was there, alone when

Emma read to him in his cigarette-fuelled afternoons, alone when they took an afternoon walk right on schedule, alone when she helped correct the grammar and spelling in his manuscripts – alone with the work that made Charles conclude that Christian accounts of eternal punishment for non-believers were grotesque. At this point in the autobiography Darwin displays a true writer's sense of phrasing and punctuation with a one-sentence paragraph:

And this is a damnable doctrine.

Six words that still convey real anger, real conviction. But also six words of final separation of man and wife. Emma's part in Darwin's life as a writer is fully displayed here. She both annotated the passage that culminates in these words *and* asked that they not be published, in the winter of 1882. Her years of living with Darwin must have prepared her for them and yet as the perfect wife she both writes them out in her own hand and asks that they be deleted after his death. And we know that Darwin felt agony about Emma's distress.

But to return to the matter of his truthfulness: if, in spite of our modern ear and its practised doubts, his words about Emma's perfection and her being Darwin's greatest blessing ring true, then it is equally true that his considered judgement on Christianity is in that one sentence. These six words are, without exaggeration, the Victorian crisis of faith in miniature. They lie on the cusp of orthodox Christianity and secular science. In his *The Descent of Man* of 1871, Darwin had described both the ways and the means by which men had achieved higher things than women in all aspects of human intellectual and practical life. When discussing Christianity and the 'damnable doctrine' in the autobiography, it is striking that the unbelievers who will suffer if that doctrine be true are all men – Darwin's father, his brother and 'almost all' his 'best friends', by whom he certainly means male friends. So these six words, when contextualized, raise a far larger question than at first appears. They raise the matter of whether women were capable of the highest of things: of scientific reasoning, of being as 'strong' as men in that regard; of whether the feminine virtues – kindness, domesticity, religious belief – were no doubt admirable and gentle but also antiquated. Nineteenth-century biomedicine had many gendered accounts of the different faculties and capacities distinguishing men and women. Darwin's particular version raises the specific matter as to whether women were strong enough to be scientifically knowledgeable religious agnostics. To face the reality that death was the end. In 1885 Leonard Darwin told his brother William that he felt their father would rather have burnt his hand off than written an autobiography which caused their mother much pain or caused dissent in the family. Yet write it Darwin did, and those words in particular. They stand between husband and wife, a mere sliver, and yet they brought with them the certainty – for both Emma and Charles Darwin – that after his death in 1882 they would never meet again.

FURTHER READING

Bowlby, John, *Charles Darwin: A New Life* (London: W. W. Norton, 1990)

Browne, Janet, *Charles Darwin: Volume 1* (London: Jonathan Cape, 1995)

Browne, Janet, *Charles Darwin: Volume 2* (London: Jonathan Cape, 2002)

Browne, Janet and Neve, Michael (eds), *Voyage of the* Beagle*: Charles Darwin's Journal of Researches* (Harmondsworth: Penguin Classics, 1989)

Colp, Ralph, Jr, 'Notes on Charles Darwin's *Autobiography*', *Journal of the History of Biology*, 1985, **18**, 357–401

Desmond, Adrian and Moore, James R., *Darwin* (Harmondsworth: Penguin, 1992)

Nicolson, Malcolm and Wilson, Jason (eds), *Personal Narrative/Alexander Von Humboldt* (Harmondsworth: Penguin Classics, 1995)

Secord, James A. (ed.), *Principles of Geology/Charles Lyell* (Harmondsworth: Penguin Classics, 1997)

Sturrock, John, *The Language of Autobiography: Studies in the First Person Singular* (Cambridge: Cambridge University Press, 1993)

NOTE ON THE TEXTS

The first of the pieces collected here, the short autobiographical 'Fragment' of August 1838, was published by John Murray in F. Darwin and A. Seward (eds), *More Letters of Charles Darwin*, vol. 1, pp. 1–5 of 1903. In 1986 it was re-edited and then printed in the Cambridge University Press series of Darwin's letters: Frederick Burkhardt, Sydney Smith et al. (eds), *The Correspondence of Charles Darwin*, vol. 2, 1837–1843, pp. 438–42. The version here is the one from the CUP volume; the meticulous list of manuscript notes and alterations introduced by the team of Cambridge editors has been removed, as have the endnotes.

The longer section, the 'Recollections' of 1876, with main additions dating from 1878 and 1881, is kept as DAR 26 in the manuscripts collection in Cambridge University Library. As discussed in the Introduction, Charles Darwin did not pen these memoirs with an eye to publication. He 'took no pains' about his writing style or the way he added material or made corrections. The chief editorial task for anyone dealing with Darwin's autobiographical writings is to 'tidy up' his thoughts. He stops, he starts, he corrects, he inserts, he stops again. Nonetheless the manuscript in its main sections flows very clearly and only at important points have the present editors indicated where additions are introduced and where they end.

Although not meant for the public eye, Darwin was very careful to signpost the exact places in the manuscript where additions were to be placed. At the point where these additions begin the editors have inserted the symbol ¶; at the point where they end they have inserted the symbol Ω. On occasion, Darwin dates his additions and these have been footnoted. His later additions consisted largely of descriptions of the character of his physician father, the character of the captain of the *Beagle*, Robert FitzRoy, and the characters of famous contemporaries whom Darwin had met. He also deleted and, less often, added words: these have also been footnoted.

We have modernized book titles and italicized Latin names but we have retained Darwin's underlinings and some of his spellings (e.g. 'Leap-year'). Other spellings have been slightly altered: this edition has 'FitzRoy' rather than 'Fitz-Roy', for example. Misspellings, such as the surname of the anthropologist E. B. Tylor appearing as 'Tyler', have been retained. Likewise, where Darwin writes 'staid' this has been kept.

The editing and censoring of the 'Recollections' from their first publication in 1887 ended in 1958 with an edition by Darwin's grand-daughter Nora Barlow. This Penguin Classics edition hopes to be truer to the manuscript in matters that might seem trivial – the use of a capital or lower case when Darwin writes 'Father' or 'father', or 'Father-Confessor', or naming the botanist Robert Brown as 'facile princeps botanicorum' at one point and 'facile Princeps Botanicorum' at another, the running on of paragraphs and so on. We have also translated some of the Latin phrases that Darwin used and footnoted these.

An autobiographical fragment

Life. Written August—1838

My earliest recollection, the date of which I can approximately tell, and which must have been before I four years old, was when sitting on Carolines knee in the dining room, whilst she was cutting an orange for me, a cow run by the window, which made me jump; so that I received a bad cut of which I bear the scar to this day. Of this scene I recollect the place where I sat & the cause of the fright, but not the cut itself.—& I think my memory is real, & not as often happens in similar cases, from hearing the thing so often repeated, one obtains so vivid an image, that it cannot be separated from memory, because I clearly remember which way the cow ran, which would not probably have been told me. My memory here is an obscure picture, in which from not recollecting any pain I am scarcely conscious of its reference to myself.—

1813 summer.—When I was four year & a half old went the sea & staid there some weeks—I remember many things, but with the exception of the maid servants (& these are not individualised) I recollect none of my family, who were there.—I remember either myself or Catherine being naughty, & being shut up in a room & trying to break the windows.—I have obscure picture of house before my eyes, & of a neighbouring small shop, where the owner gave me one fig, but which to my great joy turned out to be two:—this fig was given me that this man might kiss the maidservant:—I remember a common walk to a kind of well, on the road to which was a cottage shaded with damascene trees, inhabited by old man, called a hermit, with white hair, used to give us damascenes—I know not whether the damascenes, or the reverence &

1

indistinct fear for this old man produced the greatest effect, on my memory.—I remember, when going there crossing in the carriage a broad ford, & fear & astonishment of white foaming water has made vivid impression.—I think memory of events commences abruptly, that is I remember these *ear*liest things quite as clearly as others very much later in life, which were equally impressed on me.—Some very early recollections are connected with fear, at Parkfields with poor Betty Harvey I remember with horror her story of people being pushed into the canal by the towing rope, by going wrong side of the horse . . —I had greatest horror of this story.—keen instinct against death.—Some other recollections are those of vanity, & what is odder a consciousness, as if instinctive, & contempt of myself that I was vain—namely thinking that people were admiring me in one instance for perseverance & another for boldness in climbing a low tree.—My supposed admirer was old Peter Hailes the bricklayer, & the tree the Mountain Ash on the lawn.

All my recollections seem to be connected most closely with self.—now Catherine seems to recollect scenes, where others were chief actors.—When my mother died, I was 8 & ½ old.—& she one year less, yet she remember all particular & events of each day, whilst I scarcely recollect anything, except being sent for—memory of going into her room, my Father meeting us crying afterwards.—She remembers my mother crying, when she heard of my grandmother's death.—Also when at Parkfields, how Aunt Sarah & Kitty used to receive her—& so with very many other cases.

Susan like me, only remember affairs personal—It is sufficiently odd, this difference in subjects remembered. Catherine says she does not remember the impression made upon her by external things as scenery., but things which she reads she has excellent memory—ie for *ideas*. now her sympathy being ideal, it is part of her character, & shows how early her kind of memory was stamped. A vivid thought is repeated, a vivid impression forgotten. —

I recollect my mother's gown & scarcely anything of her appearance. except one or two walks with her I have no distinct remembrance of any conversations, & those only of very trivial nature.—I remember her saying "if she did ask me to do something, which I said she had, it was solely for my good.".—

I remember obscurely the illumination after the Battle of Waterloo, & the militia exercising, about that period, in the field opposite our House.—

1817. 8½ old went to Mr Cases school.—I remember how very much I was afraid of meeting the dogs in Barker St & how at school I could not get up my courage to fight.—I was very timid by nature. I remember I took great delight at school in fishing for newts in the quary pool—I had thus young formed a strong taste for collecting, chiefly seals, franks & but also pebbles & minerals,—one which was given me by some boy, decided this taste.—I believe shortly after this or before I had smattered in botany, & certainly when at Mr Case's school I was very fond of gardening, & invented some great falsehoods about being able to colour crocuses as I liked.—At this time I felt strong friendships for some boys.—It was soon after I began collecting stones, ie when 9 or 10 I distinctly recollect the desire I had of being able to know something about every pebble in front of the Hall door—it was my earliest—only geological aspiration at that time.—I was in these days a very great story teller,—for the pure pleasure of exciting attention & surprise. I stole fruit & hid it for these same motives, & injured trees by barking them for similar ends.—I scarcely ever went out walking without saying I had seen a pheasant or some strange bird, (natural History taste). these lies, when not detected, I presume excited my attention, as I recollect them vividly,.—not connected with shame, though some I do,—but as something which by having produced great effect on my mind, gave pleasure, like a tragedy.—

I recollect when at Mr Cases, inventing a whole fabric to show how fond I was of speaking the **truth**!—my invention is still so vivid in my mind, that I could almost fancy it was true did not memory of former shame tell me it was false.—I have no particularly happy or unhappy recollections of this time or earlier periods of my life.—

I remember well a walk I took with boy named Ford across some fields to a farmhouse on Church Stretton Road.—

I do not remember any mental pursuits excepting those of collecting stones &c.—gardening, & about this time often going with my father in his carriage, telling him of my lessons, & seeing game & other wild birds, which was a great delight to me.—I was born a naturalist.—

When I was 9 & ½ years old (July 1818) I went with Erasmus to see Liverpool.—it has left no impression in my mind, except most trifling ones.—fear of the coach upsetting, a good dinner, & an extremely vague memory of ships.

In midsummer of this year I went to D.ʳ Butlers school.—I well recollect the first going there, which oddly enough I cannot of going to Mʳ Cases, the first school of all.—I remember the year 1818 well, not from having first gone to a Public school, but from writing those figures in my school book, accompanied with obscure thoughts, now fullfilled, whether I should recollect in future life that year.—

In September (1818) I was ill with the Scarlet Fever I well remember the wretched feeling of being delirious.—

1819. July. (10 & ½ years old) Went to sea at Plas Edwards & staid there three weeks, which now appears to me like three months.—I remember a certain shady green road (where I saw a snake) & a waterfall with a degree of pleasure, which must be connected with the pleasure from scenery, though not directly recognized as such.—The sandy plain before the house has left a strong impression, which is obscurely connected with indistinct remembrance of curious insects—probably a Cimex mottled with red—the Zygena.—I was at that time very passionate, (when I swore like a trooper) & quarrelsome,—the former passion has I think nearly wholly, but slowly died away.—When journeying there by stage Coach I remember a recruiting officer (I think I should know his face to this day) at tea time, asking the maid servant for *toasted* bread butter.—I was convulsed with laughter, & thought it the quaintest & wittiest speech, that ever passed from the mouth of man.—Such is wit at 10 & ½ years old.—

The memory now flashes across me, of the pleasure I had in the evening or on blowy day walking along the beach by myseelf, & seeing the gulls & cormorants wending their way home in a wild & irregular course.—Such poetic pleasures, felt so keenly in after years,, I should not have expected so early, in life.—

1820 July. Went riding tour (on old Dobbin) with Erasmus to Pistol Rhyadwr.—of this I recollect little.—an indistinct picture of the fall.—but I well remember my astonishment on hearing that fishes could jump up it.—

(DAR 91:56–62)

NOTE ON THE AUTOBIOGRAPHICAL FRAGMENT

Darwin's family lived at The Mount, near Shrewsbury. Darwin's earliest recollection, as recorded here, involved sitting on the knee of his sister Caroline and being cut accidentally by her. He also recollected being naughty with another sister, Catherine. Sarah and Kitty were both Wedgwoods and Darwin's aunts. Parkfields was the home of Sarah Wedgwood, wife of Josiah Wedgwood I and thus Darwin's grandmother. Erasmus was Darwin's elder brother. The 'old man, called a hermit, with white hair', who donated damascenes to the children, was giving them a species of plum.

1876 May 31 – Recollections of the Development of my Mind and Character

C. Darwin

A German editor having written to me to ask for an account of the development of my mind and character with some sketch of my autobiography, I have thought that the attempt would amuse me, and might possibly interest my children or their children. I know that it would have interested me greatly to have read even so short and dull a sketch of the mind of my grandfather written by himself, and what he thought and did and how he worked. I have attempted to write the following account of myself, as if I were a dead man in another world looking back at my own life. Nor have I found this difficult, for life is nearly over with me. I have taken no pains about my style of writing.

I was born at Shrewsbury on February 12th, 1809. I have heard my father say that he believed that persons with powerful minds generally had memories extending far back to a very early period of life. This is not my case for my earliest recollection goes back only to when I was a few months over four years old, when we went to near Abergele for sea-bathing, and I recollect some events and places there with some little distinctness.

My mother died in July 1817, when I was a little over eight years old, and it is odd that I can remember hardly anything about her except her death-bed, her black velvet gown, and her curiously constructed work-table. I believe that my forgetfulness is partly due to my sisters, owing to their great grief, never being able to speak about her or mention her name; and partly to her previous invalid state. In the spring of this same year I was sent to a day-school in Shrewsbury, where I staid a year. Before going to school I was educated by my sister Caroline, but

I doubt whether this plan answered. I have been told that I was much slower in learning than my younger sister Catherine, and I believe that I was in many ways a naughty boy. Caroline was extremely kind, clever and zealous; but she was too zealous in trying to improve me; for I clearly remember after this long interval of years, saying to myself when about to enter a room where she was – "What will she blame me for now?" and I made myself dogged so as not to care what she might say.

By the time I went to this day-school my taste for natural history, and more especially for collecting, was well developed. I tried to make out the names of plants and collected all sorts of things, shells, seals, franks, coins, and minerals. The passion for collecting, which leads a man to be a systematic naturalist, a virtuoso or a miser, was very strong in me, and was clearly* innate, as none of my sisters or brother ever had this taste.

One little event during this year has fixed itself very firmly in my mind, and I hope that it has done so from my conscience having been afterwards sorely troubled by it; it is curious as showing that apparently I was interested at this early age in the variability of plants! I told another little boy (I believe it was Leighton, who afterwards become a well-known Lichenologist and botanist) that I could produce variously coloured Polyanthuses and Primroses by watering them with certain coloured fluids, which was of course a monstrous fable, and had never been tried by me. I may here also confess that as a little boy I was much given to inventing deliberate falsehoods, and this was always done for the sake of causing excitement. For instance, I once gathered much valuable fruit from my Father's trees and hid them in the shrubbery, and then ran in breathless haste to spread the news that I had discovered a hoard of stolen fruit.

¶ About this time, or as I hope at a somewhat earlier age, I sometimes stole fruit for the sake of eating it; and one of my schemes was ingenious. The kitchen garden was kept locked in the evening, and was surrounded by a high wall, but by the aid of neighbouring trees I could easily get

* 'inherent' deleted.

on the coping. I then fixed a long stick into the hole at the bottom of a rather large flower-pot, and by dragging this upwards pulled off peaches and plums, which fell into the pot and the prizes were thus secured. When a very little boy I remember stealing apples from the orchard, for the sake of giving them away to some boys and young men who lived in a cottage not far off, but before I gave them the fruit I showed off how quickly I could run and it is wonderful that I did not perceive that the surprise and admiration which they expressed at my powers of running was given for the sake of the apples. But I well remember that I was delighted at them declaring that they had never seen a boy run so fast!

I remember clearly only one other incident during the years whilst at Mr. Case's daily school – namely, the burial of a dragoon-soldier; and it is surprising how clearly I can still see the horse with the man's empty boots and carbine suspended to the saddle, and the firing over the grave. This scene deeply stirred whatever poetic fancy there was in me. Ω

In the summer of 1818 I went to Dr. Butler's great school in Shrewsbury, and remained there for seven years till Mid-summer 1825, when I was sixteen years old. I boarded at this school, so that I had the great advantage of living the life of a true school-boy; but as the distance was hardly more than a mile to my home, I very often ran there in the longer intervals between the callings over and before locking up at night. This I think was in many ways advantageous to me by keeping up home affections and interests. I remember in the early part of my school life that I often had to run very quickly to be in time, and from being a fleet runner was generally successful; but when in doubt I prayed earnestly to God to help me, and I well remember that I attributed my success to the prayers and not to my quick running, and marvelled how generally I was aided.

I have heard my father and elder sisters say that I had, as a very young boy, a strong taste for long solitary walks; but what I thought about I know not. I often became quite absorbed, and once, whilst returning to school on the summit of the old fortifications round Shrewsbury, which had been converted into a public foot-path with no

8

parapet on one side, I walked off and fell to the ground, but the height was only seven or eight feet. Nevertheless the number of thoughts which passed through my mind during this very short, but sudden and wholly unexpected fall, was astonishing, and seem hardly compatible with what physiologists have, I believe, proved about each thought requiring quite an appreciable amount of time.

¶ I must have been a very* simple little fellow when I first went to the school. A boy of the name of Garnett took me into a cake-shop one day, and bought some cakes for which he did not pay, as the shopman trusted him. When we came out I asked him why he did not pay for them, and he instantly answered, "Why, do you not know that my uncle left a great sum of money to the Town on condition that every tradesman should give whatever was wanted without payment to anyone who wore his old hat and moved it in a particular manner;" and he then showed me how it was moved. He then went into another shop where he was trusted, and asked for some small article, moving his hat in the proper manner, and of course obtained it without payment. When we came out he said, "Now if you like to go by yourself into that cake-shop (how well I remember its exact position), I will lend you my hat, and you can get whatever you like if you move the hat on your head properly." I gladly accepted the generous offer, and went in and asked for some cakes, moved the old hat, and was walking out of the shop, when the shopman made a rush at me, so I dropped the cakes and ran away for dear life, and was astonished by being greeted with shouts of laughter by my false friend Garnett.

I can say in my own favour that I was as a boy humane, but I owed this entirely to the instruction and example of my sisters. I doubt indeed whether humanity is a natural or innate quality. I was very fond of collecting eggs, but I never took more than a single egg out of a bird's nest, except on one single occasion, when I took all, not for their value, but from a sort of bravado.

I had a strong taste for angling, and would sit for any number of hours on the bank of a river or pond watching the float; when at Maer

* 'wonderfully' deleted.

I was told that I could kill the worms with salt and water, and from that day I never spitted a living worm, though at the expense, as I believe, of some loss of success.

Once as a very little boy, whilst at the day-school, or before that time, I acted cruelly, for I beat a puppy I believe, simply from enjoying the sense of power; but the beating could not have been severe, for the puppy did not howl, of which I feel sure as the spot was near to the house. This act lay heavily on my conscience, as is shown by my remembering the exact spot where the crime was committed. It probably* lay all the heavier from my love of dogs being then, and for a long time afterwards, a passion. Dogs seemed to know this, for I was an adept in robbing their love from their masters. Ω

Nothing could have been worse for the development of my mind than Dr. Butler's school, as it was strictly classical, nothing else being taught except a little ancient geography and history. The school as a means of education to me was simply a blank. During my whole life I have been singularly incapable of mastering any language. Especial attention was paid to verse-making, and this I could never do well. I had many friends, and got together a grand collection of old verses, which by patching together, sometimes aided by other boys, I could work into any subject. Much attention was paid to learning by heart the lessons of the previous day; this I could effect with great facility learning 40 or 50 lines of Virgil or Homer, whilst I was in morning chapel; but this exercise was utterly useless, for every verse was forgotten in 48 hours. I was not idle, and with the exception of versification, generally worked conscientiously at my classics, not using cribs. The sole pleasure I ever received from such studies, was from some of the odes of Horace, which I admired greatly. When I left the school I was for my age neither high nor low in it; and I believe that I was considered by all my masters and by my Father as a very ordinary boy, rather below the common standard in intellect. To my deep mortification my father once said to me, "You care for nothing but shooting, dogs, and rat-catching, and you will be a disgrace to yourself and all your family." But my father, who was the kindest man

* 'probably' added.

I ever knew, and whose memory I love with all my heart, must have been angry and somewhat unjust when he used such words.

¶ I may here add a few pages about my Father, who was in many ways a remarkable man. He was about 6 ft 2 inches in height, with broad shoulders, and very corpulent, so that he was the largest man whom I ever saw. When he last weighed himself he was 24 stone, but afterwards increased much in weight. His chief mental characteristics were his powers of observation and his sympathy, neither of which have I ever seen exceeded or even equalled. His sympathy was not only with the distresses of others, but in a greater degree with the pleasures of all around him. This led him to be always scheming to give pleasure to others, and, though hating extravagance, to perform many generous actions. For instance, Mr B a small manufacturer in Shrewsbury, came to him one day, and said he should be bankrupt unless he could at once borrow £10,000, but that he was unable to give any legal security. My Father heard his reasons for believing that he could ultimately repay the money, and from my father's intuitive perception of character felt sure that he was to be trusted. So he advanced this sum, which was a very large one for him while young, and was after a time repaid.

I suppose that it was his sympathy which gave him unbounded power of winning confidence, and as a consequence made him highly successful as a physician. He began to practise before he was twenty-one years old, and his fees during the first year paid for the keep of two horses and a servant. On the following year his practice was larger, and so continued for above 60 years, when he ceased to attend on any one. His great success as a doctor was the more remarkable, as he told me that he at first hated his profession so much that if he had been sure of the smallest pittance, or if his father had given him any choice, nothing should have induced him to follow it. To the end of his life, the thought of an operation almost sickened him, and he could scarcely endure to see a person bled – a horror which he has transmitted to me – and I remember the horror which I felt as a schoolboy in reading about Pliny (I think) bleeding to death in a warm bath. My father told me two odd stories about bleeding: one was that as a very young man he became a Freemason. A friend of his who was a freemason and who pretended not to know about his strong feeling with respect to blood, remarked casually to him as they walked to the meeting, "I suppose that you do not care about losing a few drops of blood?" It seems that when he was received as a member, his eyes were bandaged and his coat-sleeves turned up. Whether any such ceremony is now performed I know not, but my Father mentioned the case as an excellent instance of the power of imagination, for he distinctly felt the blood trickling down his arm, and could hardly believe his own eyes, when he afterwards could not find the smallest prick on his arm.

A great slaughtering butcher from London once consulted my grandfather, when another man very ill was brought in, and my grandfather wished to have him instantly bled by the accompanying apothecary. The butcher was asked to hold the patient's arm, but he made some excuse and left the room. Afterwards he explained to my grandfather that although he believed that he had killed with his own hands more animals than any other man in London, yet absurd as it might seem he assuredly should have fainted if he had seen the patient bled.

Owing to my father's power of winning confidence, many patients, especially ladies, consulted him when suffering from any misery, as a sort of Father-Confessor. He told me that they always began by complaining in a vague manner about their health, and by practice he soon guessed what was really the matter. He then suggested that they had been suffering in their minds, and now they would pour out their troubles, and he heard nothing more about the body. Family quarrels were a common subject. When gentlemen complained to him about their wives, and the quarrel seemed serious, my Father advised them to act in the following manner; and his advice always succeeded if the gentleman followed it to the letter, which was not always the case. The husband was to say to the wife that he was very sorry that they could not live happily together,—that he felt sure that she would be happier if separated from him – that he did not blame her in the least (this was the point on which the man oftenest failed) – that he would not blame her to any of her relations or friends and lastly that he would settle on her as large a provision as he could afford. She was then asked to deliberate on this proposal. As no fault had been found, her temper was unruffled, and she soon felt what an awkward position she would be in, with no accusation to rebut, and with her husband and not herself proposing a separation. Invariably the lady begged her husband not to think of separation, and usually behaved much better ever afterwards.

Owing to my father's skill in winning confidence he received many strange confessions of misery and guilt. He often remarked how many miserable wives he had known. In several instances husbands and wives had gone on pretty well together for between 20 and 30 years, and then hated each other bitterly: this he attributed to their having lost a common bond in their young children having grown up.

But the most remarkable power which my father possessed was that of reading the characters, and even the thoughts of those whom he saw even for a short time. We had many instances of this power, some of which seemed almost supernatural. It saved my father from ever making (with one exception, and the character of this man was soon discovered) an unworthy friend. A strange clergyman came to Shrewsbury, and seemed to be a rich man; everybody called on him, and he was invited to many houses. My father called, and on his return home told my sisters on no account to invite him or his family to our house; for he felt sure that the man was not to be trusted. After a few months he suddenly bolted, being heavily in debt, and was found out to be little better than an habitual swindler. Here is a case of trustfulness which not many men would have ventured on. An Irish gentleman, a complete stranger, called on my father one day, and said that he had lost his purse, and that it would be a serious inconvenience to him to wait in Shrewsbury until he could receive a remittance from Ireland. He then asked my father to lend him £20, which was immediately done, as my father felt certain that the story was a true one. As soon as a letter could arrive from Ireland, one came with the most profuse thanks, and enclosing, as he said, a £20 Bank of England note; but no note was enclosed. I asked my father whether this did not stagger him, but he answered "not in the least." On the next day another letter came with many apologies for having forgotten (like a true Irishman) to put the note into his letter of the day before.

A connection of my Father's consulted him about his son who was strangely idle and would settle to no work. My father said "I believe that the foolish young man thinks that I shall bequeath him a large sum of money. Tell him that I have declared to you that I shall not leave him a penny." The father of the youth owned with shame that this preposterous idea had taken possession of his son's mind; and he asked my father how he could possibly have discovered it, but my father said he did not in the least know.

The Earl of —— brought his nephew, who was insane but quite gently, to my father; and the young man's insanity led him to accuse himself of all the crimes under heaven. When my father afterwards talked about the case with the uncle, he said, "I am sure that your nephew is really guilty of . . . a heinous crime." Whereupon the Earl of —— exclaimed, "Good God, Dr. Darwin, who told you; we thought that no human being knew the fact except ourselves!" My father told me the story many years after the event, and I asked him how he distinguished the true from the false self-accusations; and it was very characteristic of my Father that he said he could not explain how it was.

The following story shows what good guesses my father could make. Lord Sherburn, afterwards the first Marquis of Lansdowne, was famous (as Macaulay somewhere remarks) for his knowledge of the affairs of Europe, on which he greatly prided himself. He consulted my father medically, and afterwards harangued him on the state of Holland. My father had studied medicine at Leyden, and one day went a long walk into the country with a friend, who took him to the house of a clergyman (we will say the Rev. Mr A——, for I have forgotten his name), who had married an Englishwoman. My father was very hungry, and there was little for luncheon except cheese, which he could never eat. The old lady was surprised and grieved at this, and assured my father that it was an excellent cheese, and had been sent her from Bowood, the seat of Lord Sherburn. My father wondered why a cheese should be sent her from Bowood, but thought nothing more about it until it flashed across his mind many years afterwards, whilst Lord Sherburn was talking about Holland. So he answered, "I should think from what I saw of the Rev. Mr A——, that he was a very able man and well acquainted with the state of Holland." My father saw that the Earl, who immediately changed the conversation, was much startled. On the next morning my father received a note from the Earl, saying that he had delayed starting on his journey, and wished particularly to see my father. When he called, the Earl said, "Dr. Darwin, it is of the utmost importance to me and to the Rev. Mr A —— to learn how you have discovered that he is the source of my information about Holland." So my father had to explain the state of the case, and he supposed that Lord Sherburn was much struck with his diplomatic skill in guessing, for during many years afterwards he received many kind messages from him through various friends. I think that he must have told the story to his children; for Sir C. Lyell asked me many years ago why the Marquis of Lansdowne (the son or grandson of the first marquis) felt so much interest about me, whom he had never seen, and my family. When 40 new members (the 40 thieves as they were then called) were added to the Athenæum Club, there was much canvassing to be one of them; and without my having asked any one, Lord Lansdowne proposed me and got me elected. If I am right in my supposition, it was a queer concatenation of events that my father not eating cheese half-a-century before in Holland led to my election as a member of the Athenæum.

Early in life my father occasionally wrote down a short account of some curious event and conversation, which are enclosed in a separate envelope.

The sharpness of his observation led him to predict with remarkable skill the course of any illness, and he suggested endless small details of relief. I was told that a young Doctor in Shrewsbury, who disliked my father, used to say that he was wholly unscientific, but owned that his power of predicting the end of an illness was unparalleled. Formerly when he thought that I should be a doctor, he talked much to me about his patients. In the old days the practice of bleeding largely was universal, but my father maintained that far more evil was thus caused than good done; and he advised me if ever I was myself ill not to allow any doctor to take from me more than an extremely small quantity of blood. Long before typhoid fever was recognised as distinct, my father told me that two utterly distinct kinds of illness were confounded under the name of typhus fever. He was vehement against drinking, and was convinced of both the direct and inherited evil effects of alcohol when habitually taken even in moderate quantity in a very large majority of cases. But he admitted and advanced instances of certain persons, who could drink largely during their whole lives without apparently suffering any evil effects; and he believed that he could often beforehand tell who would thus not suffer. He himself never drank a drop of any alcoholic fluid.

This remark reminds me of a case showing how a witness under the most favourable circumstances may be wholly mistaken. A gentleman-farmer was strongly urged by my father not to drink, and was encouraged by being told that he himself never touched any spirituous liquor. Whereupon the gentleman said, "Come, come, Doctor, that won't do – though it is very kind of you to say so for my sake – for I know that you take a very large glass of hot gin and water every evening after your dinner." So my father asked him how he knew this. The man answered, "My cook was your kitchen-maid for two or three years, and she saw the butler every day prepare and take to you the gin and water." The explanation was that my father had the odd habit of drinking hot water in a very tall and large glass after his dinner; and the butler used first to put some cold water in the glass, which the girl mistook for gin, and then filled it up with boiling water from the kitchen-boiler.Ω

My father used to tell me many little things which he had found useful in his medical practice. Thus ladies often cried much while telling him their troubles, and thus caused much loss of his precious time. He soon found that begging them to command and restrain themselves, always made them weep the more, so that afterwards he always encouraged them to go on crying, saying that this would relieve them more than anything else, with the invariable result that they soon ceased to cry, and he could hear what they had to say and give his advice. When patients who were very ill, craved for some strange and unnatural food, my father asked them what had put such an idea into their heads: if they answered that they did not know, he would allow them to try the food, and often with success, as he trusted to their having a kind of instinctive desire; but if they answered that they had heard that the food in question had done good to someone else, he firmly refused his assent.

He gave one day an odd little specimen of human nature. When a very young man he was called in to consult with the family physician in the case of a gentleman of much distinction in Shropshire. The old doctor told the wife that the illness was of such a nature that it must end fatally. My father took a different view and maintained that the gentleman would recover: he was proved quite wrong in all respects, (I think by autopsy) and he owned his error. He was then convinced that he should never again be consulted by this family; but after a few months the widow sent for him, having dismissed the old family doctor. My father was so much surprised at this, that he asked a friend of the widow to find out why he was again consulted. The widow answered her friend, that "she would never again see that odious old doctor who said from the first that her husband would die, while Dr. Darwin always maintained that he would recover!" In another case my father told a lady that her husband would certainly die. Some months afterwards he saw the widow who was a very sensible woman, and she said, "You are a very young man, and allow me to advise you always to give, as long as you possibly can, hope to any near relation nursing a patient. You made me despair, and from that moment I lost strength." My father said that he had often since seen the paramount importance, for the sake of the patient, of keeping up the hope and with it the strength of the nurse in charge. This he sometimes found it difficult to do compatibly with truth. One old gentleman, however, Mr. Pemberton, caused him no such perplexity. He was sent for by Mr. Pemberton, who said, "From all that I have seen and heard of you I believe you are the sort of man who will speak the truth, and if I ask you will tell me when I am dying. Now I much desire that you should attend me, if you will promise, whatever I may say, always to declare that I am not going to die." My father acquiesced on this understanding that his words should in fact have no meaning.

My father possessed an extraordinary memory, especially for dates, so that he knew, when he was very old the day of the birth, marriage, and death of a multitude of persons in Shropshire; and he once told me that this power annoyed him; for if he once heard a date he could not forget it; and thus the deaths of many friends were often recalled to his mind. Owing to his strong memory he knew an extraordinary number of curious stories, which he liked to tell, as he was a great talker. He was generally in high spirits, and laughed and joked with every one – often with his servants – with the utmost freedom; yet he had the art of making every one obey him to the letter. Many persons were much afraid of him. I remember my father telling us one day with a laugh, that several persons had asked him whether Miss Piggott (a grand old lady in Shropshire), had called on him, so that at last he enquired why they asked him; and was told that Miss Piggott, whom my father had somehow mortally offended, was telling everybody that she would call and tell 'that fat old doctor very plainly what she thought of him.' She had already called, but her courage had failed, and no one could have been more courteous and friendly. As a boy, I went to stay at the house of Major B——, whose wife was insane; and the poor creature, as soon as she saw me, was in the most abject state of terror that I ever saw, weeping bitterly and asking me over and over again, "Is your father coming?" but was soon pacified. On my return home, I asked my father why she was so frightened, and he answered he was very glad to hear it, as he had frightened her on purpose, feeling sure that she could be kept in safety and much happier without any restraint, if her husband could influence her, whenever she became at all violent, by proposing to send for Dr. Darwin; and these words succeeded perfectly during the rest of her long life.

My father was very sensitive so that many small events annoyed or pained him much. I once asked him, when he was old and could not walk, why he did not drive out for exercise; and he answered, "Every road out of Shrewsbury is associated in my mind with some painful event." Yet he was generally in high spirits. He was easily made very angry, but as his kindness was unbounded, he was widely and deeply loved.

He was a cautious and good man of business, so that he hardly ever lost money by any investment, and left to his children a very large property. I remember a story, showing how easily utterly false beliefs originate and spread. Mr E——, a squire of one of the oldest families in Shropshire, and head partner in a Bank, committed suicide. My father was sent for as a matter of form, and found him dead. I may mention by the way, to show how matters were managed in those old days, that because Mr E—— was a rather great man and universally respected, no inquest was held over his body. My father, in returning home, thought it proper to call at the Bank (where he had an account) to tell the managing partner of the event, as it was not improbable it would cause a run on the bank. Well the story was spread far and wide, that my father went into the bank, drew out all his money, left the bank, came back again, and said, "I may just tell you that Mr E—— has killed himself," and then departed. It seems that it was then a common belief that money withdrawn from a bank was not safe, until the person had passed out through the door of the bank. My father did not hear this story till some little time afterwards, when the managing partner said that he had departed from his invariable rule of never allowing any one to see the account of another man, by having shown the ledger with my father's account to several persons, as this proved that my father had not drawn out a penny on that day. It would have been dishonourable in my father to have used his professional knowledge for his private advantage. Nevertheless the supposed act was greatly admired by some persons; and many years afterwards, a gentleman remarked, "Ah, Doctor, what a splendid man of business you were in so cleverly getting all your money safe out of that bank."

My father's mind was not scientific, and he did not try to generalise his knowledge under general laws; yet he formed a theory for almost everything which occurred. I do not think that I gained much from him intellectually, but his example ought to have been of much moral service to all his children. One of his golden rules (a hard one to follow) was, "Never become the friend of any one whom you cannot respect."

With respect to my father's father, the author of the *Botanic Garden* etc., I have put together all the facts which I could collect in his published *Life*.

Having said this much about my father, I will add a few words about my brother and sisters. My brother Erasmus possessed a remarkably clear mind, with extensive and diversified tastes and knowledge in literature, art, and even in science. For a short time he collected and dried plants, and during a somewhat longer time experimented in chemistry. He was extremely agreeable, and his wit often reminded me of that in the letters and works of Charles Lamb. He was very kind-hearted; but his health from his boyhood had been weak, and as a consequence he failed in energy. His spirits were not high, sometimes low, more especially during early and middle manhood. He read much, even whilst a boy, and at school encouraged me to read, lending me books. Our minds and tastes were, however, so different that I do not think that I owe much to him intellectually – nor to my four sisters, who possessed very different characters, and some of them had strongly marked characters. All were extremely kind and affectionate towards me during their whole lives. I am inclined to agree with Francis Galton in believing that education and environment produce only a small effect on the mind of any one, and that most of our qualities are innate.

The above sketch of my brother's character was written before that which was published in Carlyle's *Remembrances*, and which appears to me to have little truth and no merit.

Looking back as well as I can at my character during my school life, the only qualities which at this period promised well for the future, were, that I had strong and diversified tastes, much zeal for whatever interested me, and a keen pleasure in understanding any complex subject or thing. I was taught Euclid by a private tutor, and I distinctly remember the intense satisfaction which the clear geometrical proofs gave me. I remember with equal distinctness the delight which my uncle gave me (the father of Francis Galton) by explaining the principle of the vernier of a barometer. With respect to diversified tastes, independently of science, I was fond of reading various books, and I used to sit for hours reading the historical plays of Shakespeare, generally in an old window in the thick walls of the school. I read also other poetry, such as the recently published poems of Byron, Scott, and Thomson's *Seasons*. I mention this because later in life I wholly lost, to my great regret, all pleasure from poetry of any kind, including Shakespeare. In connection with pleasure from poetry I may add that in 1822 a vivid delight in scenery was first awakened in my mind, during a riding tour on the borders of Wales, and which has lasted longer than any other aesthetic pleasure.

Early in my school-days a boy had a copy of the *Wonders of the*

World, which I often read and disputed with other boys about the veracity of some of the statements; and I believe this book first gave me a wish to travel in remote countries, which was ultimately fulfilled by the voyage of the *Beagle*. In the latter part of my school life I became passionately fond of shooting, and I do not believe that anyone could have shown more zeal for the most holy cause than I did for shooting birds. How well I remember killing my first snipe, and my excitement was so great that I had much difficulty in reloading my gun from the trembling of my hands. This taste long continued and I became a very good shot. When at Cambridge I used to practise throwing up my gun to my shoulder before a looking-glass to see that I threw it up straight. Another and better plan was to get a friend to wave about a lighted candle, and then to fire at it with a cap on the nipple, and if the aim was accurate the little puff of air would blow out the candle. The explosion of the cap caused a sharp crack, and I was told that the Tutor of the College remarked, "What an extraordinary thing it is, Mr Darwin seems to spend hours in cracking a horse-whip in his room, for I often hear the crack when I pass under his windows."

I had many friends amongst the schoolboys, whom I loved dearly, and I think that my disposition was then very affectionate. Some of these boys were rather clever, but I may add on the principle of "noscitur a socio"* that not one of them ever became in the least distinguished.

With respect to science, I continued collecting minerals with much zeal, but quite unscientifically – all that I cared for was a new <u>named</u> mineral, and I hardly attempted to classify them. I must have observed insects with some little care, for when ten years old (1819) I went for three weeks to Plas Edwards on the sea-coast in Wales, I was very much interested and surprised at seeing a large black and scarlet Hemipterous insect, many moths (Zygæna) and a Cicindela, which are not found in Shropshire. I almost made up my mind to begin collecting all the insects which I could find dead, for on consulting my sister, I concluded that it was not right to kill insects for the sake of making a collection. From reading White's *Selborne* I took much pleasure in watching the habits

* CD uses a Latin proverb meaning that one may judge a man from the friends he keeps.

of birds, and even made notes on the subject. In my simplicity I remember wondering why every gentleman did not become an ornithologist.

Towards the close of my school-life, my brother worked hard at chemistry and made a fair laboratory with proper apparatus in the tool-house in the garden, and I was allowed to aid him as a servant in most of his experiments. He made all the gases and many compounds, and I read with care several books on chemistry, such as Henry and Parkes' *Chemical Catechism*. The subject interested me greatly, and we often used to go on working till rather late at night. This was the best part of my education at school, for it showed me practically the meaning of experimental science. The fact that we worked at chemistry somehow got known at school, and as it was an unprecedented fact, I was nick-named "Gas." I was also once publicly rebuked by the head-master, Dr. Butler, for thus wasting my time over such useless subjects; and he called me very unjustly a "poco curante," and as I did not understand what he meant it seemed to me a fearful reproach.

As I was doing no good at school, my father wisely took me away at a rather earlier age than usual, and sent me (October 1825) to Edinburgh University with my brother, where I stayed for two years or sessions. My brother was completing his medical studies, though I do not believe he ever really intended to practise, and I was sent there to commence them. But soon after this period I became convinced from various small circumstances that my father would leave me property enough to subsist on with some comfort, though I never imagined that I should be so rich a man as I am; but my belief was sufficient to check any strenuous effort to learn medicine.

The instruction at Edinburgh was altogether by Lectures, and these were intolerably dull, with the exception of those on chemistry by Hope; but to my mind there are no advantages and many disadvantages in lectures compared with reading. Dr. Duncan's lectures on Materia Medica at 8 o'clock on a winter's morning are something fearful to remember. Dr. Munro made his lectures on human anatomy as dull, as he was himself, and the subject disgusted me. It has proved one of the greatest evils in my life that I was not urged to practise dissection, for I should soon have got over my disgust; and the practice would have been invaluable for all my future work. This has been an irremediable evil, as well as my incapacity to draw. I also attended regularly the clinical wards in the Hospital. Some of the cases distressed me a good deal, and I still have vivid pictures before me of some of them; but I was not so foolish as to allow this to lessen my attendance. I cannot understand why this part of my medical course did not interest me in a greater degree; for during the summer before coming to Edinburgh I began attending some of the poor people, chiefly children and women in Shrewsbury: I wrote down as full an account as I could of the cases with all the symptoms, and read them aloud to my father, who suggested further enquiries, and advised me what medicines to give, which I made up myself. At one time I had at least a dozen patients, and I felt a keen interest in the work. My father, who was by far the best judge of character whom I ever knew, declared that I should make a successful physician, – meaning by this, one who got many patients. He maintained that the chief element of success was exciting confidence; but what he saw in me which convinced him that I should create confidence I know not. I also attended on two occasions the operating theatre in the hospital at Edinburgh, and saw two very bad operations, one on a child, but I rushed away before they were completed. Nor did I ever attend again, for hardly any inducement would have been strong enough to make me do so; this being long before the blessed days of chloroform. The two cases fairly haunted me for many a long year.

My Brother staid only one year at the University, so that during the second year I was left to my own resources; and this was an advantage, for I became well acquainted with several young men fond of natural science. One of these was Ainsworth, who afterwards published his travels in Assyria: he was a Wernerian geologist and knew a little about many subjects, but was superficial and very glib with his tongue. Dr. Coldstream was a very different young man, prim, formal, highly religious and most kind-hearted: he afterwards published some good zoological articles. A third young man was Hardie, who wd. I think have made a good botanist, but died early in India. Lastly, Dr. Grant, my senior by several years, but how I became acquainted with him I cannot remember: he published some first-rate zoological papers, but after coming to London as Professor in University College, he did nothing more in science – a fact which has always been inexplicable to me. I knew him well; he was dry and formal in manner, but with much enthusiasm beneath this outer crust. He one day, when we were walking together burst forth in high admiration of Lamarck and his views on evolution. I listened in silent astonishment, and as far as I can judge, without any effect on my mind. I had previously read the *Zoonomia* of my grandfather, in which similar views are maintained, but without producing any effect on me. Nevertheless it is probable that the hearing rather early in life such views maintained and praised may have favoured my upholding them under a different form in my *Origin of Species*. At this time I admired greatly the *Zoonomia*; but on reading it a second time after an interval of ten or fifteen years, I was much disappointed, the proportion of speculation being so large to the facts given.

Drs. Grant and Coldstream attended much to marine Zoology, and I often accompanied the former to collect animals in the tidal pools, which I dissected as well as I could. I also became friends with some of the Newhaven fishermen, and sometimes accompanied them when they trawled for oysters, and thus got many specimens. But from not having had any regular practice in dissection, and from possessing only a wretched microscope my attempts were very poor. Nevertheless I made one interesting little discovery, and read about the beginning of the year 1826, a short paper on the subject before the Plinian Socy. This was that the so-called ova of Flustra had the power of independent movement by means of cilia, and were in fact larvæ. In another short paper I showed that little globular bodies which had been supposed to be the young state of *Fucus loreus* were the egg-cases of the worm-like *Pontobdella muricata*.

The Plinian Society was encouraged and I believe founded by Professor Jameson: it consisted of students and met in an underground room in the University for the sake of reading papers on natural science and discussing them. I used regularly to attend and the meetings had a good effect on me in stimulating my zeal and giving me new congenial acquaintances. One evening a poor young man got up and after stammering for a prodigious length of time, blushing crimson, he at last slowly got out the words, "Mr. President, I have forgotten what I was going to say." The poor fellow looked quite overwhelmed, and all the members were so surprised that no one could think of a word to say to cover his confusion. The papers which were read to our little society were not printed, so that I had not the satisfaction of seeing my paper in print; but I believe Dr. Grant noticed my small discovery in his excellent memoir on Flustra.

I was also a member of the Royal Medical Society, and attended pretty regularly, but as the subjects were exclusively medical I did not much care about them. Much rubbish was talked there, but there were some good speakers, of whom the best was the present Sir J. Kay-Shuttleworth. Dr. Grant took me occasionally to the meetings of the Wernerian Society, where various papers on natural history were read, discussed, and afterwards published in the Transactions. I heard Audubon deliver there some interesting discourses on the habits of N. American birds, sneering somewhat unjustly at Waterton. By the way, a negro lived in Edinburgh, who had travelled with Waterton and gained his livelihood by stuffing birds, which he did excellently; he gave me lessons for payment, and I used often to sit with him, for he was a very pleasant and intelligent man.

Mr Leonard Horner also took me once to a meeting of the Royal Society of Edinburgh, where I saw Sir Walter Scott in the chair as President, and he apologised to the meeting as not feeling fitted for such a position. I looked at him and at the whole scene with some awe and reverence; and I think it was owing to this visit during my youth and to my having attended the Royal Medical Society, that I felt the honour of being elected a few years ago an honorary member of both these Societies, more than any other similar honour. If I had been told at that time that I should one day have been thus honoured, I declare that I should have thought it as ridiculous and improbable, as if I had been told that I should be elected King of England.

During my second year in Edinburgh I attended Jameson's lectures on Geology and Zoology, but they were incredibly dull. The sole effect they produced on me was the determination never as long as I lived to read a book on Geology or in any way to study the science. Yet I feel sure that I was prepared for a philosophical treatment of the subject; for an old Mr Cotton in Shropshire who knew a good deal about rocks, had pointed out to me, 2 or 3 years previously a well-known large erratic boulder in the town of Shrewsbury called the bell-stone; he told me that there was no rock of the same kind nearer than Cumberland or Scotland, and he solemnly assured me that the world would come to an end before anyone would be able to explain how this stone came where it now lay. This produced a deep impression on me and I meditated over this wonderful stone. So that I felt the keenest delight when I first read of the action of icebergs in transporting boulders, and I gloried in the progress of Geology. Equally striking is the fact that I, though now only sixty-seven years old, heard Professor Jameson, in a field lecture at Salisbury Craigs, discoursing on a trap-dyke, with amygdaloidal margins and the strata indurated on each side, with volcanic rocks all around us, and say that it was a fissure filled with sediment from above, adding with a sneer that there were men who maintained that it had been injected from beneath in a molten condition. When I think of this lecture, I do not wonder that I determined never to attend to Geology.

From attending Jameson's lectures, I became acquainted with the curator of the museum, Mr. Macgillivray, who afterwards published a large and excellent book on the birds of Scotland. He had not much the appearance or manners of the gentleman. I had much interesting natural-history talk with him, and he was very kind to me. He gave me some rare shells, for I at that time collected marine mollusca, but with no great zeal.

My summer vacations during these two years were wholly given up to amusements, though I always had some book in hand, which I read with interest. During the summer of 1826, I took a long walking tour with two friends with knapsacks on our backs through North Wales. We walked thirty miles most days, including one day the ascent of Snowdon. I also went with my sister Caroline a riding tour in North Wales, a servant with saddle-bags carrying our clothes. The autumns were devoted to shooting, chiefly at Mr. Owen's at Woodhouse, and at my Uncle Jos's, at Maer. My zeal was so great that I used to place my shooting boots open by my bed-side when I went to bed, so as not to lose half-a-minute in putting them on in the morning; and on one occasion I reached a distant part of the Maer estate on the 20th of August for black-game shooting, before I could see: I then toiled on with the gamekeeper the whole day through thick heath and young Scotch firs. I kept an exact record of every bird which I shot throughout the whole season.

¶ One day when shooting at Woodhouse with Captain Owen, the eldest son and Major Hill, his cousin, afterwards Lord Berwick, both of whom I liked very much, I thought myself shamefully used, for every time after I had fired and thought that I had killed a bird, one of the two acted as if loading his gun and cried out, "You must not count that bird, for I fired at the same time," and the gamekeeper perceiving the joke, backed them up. After some hours they told me the joke, but it was no joke to me for I had shot a large number of birds, but did not know how many, and could not add them to my list, which I used to do by making a knot in a piece of string tied to a button-hole. This my wicked friends had perceived. Ω

How I did enjoy shooting, but I think that I must have been half-consciously ashamed of my zeal, for I tried to persuade myself that shooting was almost an intellectual employment; it required so much skill to judge where to find most game and to hunt the dogs well.

One of my autumnal visits to Maer in 1827 was memorable from meeting there Sir J. Mackintosh, who was the best converser I ever listened to. I heard afterwards with a glow of pride that he had said, "There is something in that young man that interests me." This must have been chiefly due to his perceiving that I listened with much interest to everything which he said, for I was as ignorant as a pig about his subjects of history, politicks and moral philosophy. To hear of praise from an eminent person, though no doubt apt or certain to excite vanity, is, I think, good for a young man, as it helps to keep him in the right course.

My visits to Maer during these two and the three succeeding years were quite delightful, independently of the autumnal shooting. Life there was perfectly free; the country was very pleasant for walking or riding; and in the evening there was much very agreeable conversation, not so personal as it generally is in large family parties, together with music. In the summer the whole family used often to sit on the steps of the old portico, with the flower-garden in front, and with the steep wooded bank, opposite the house, reflected in the lake, with here and there a fish rising or a water-bird paddling about. Nothing has left a more vivid picture on my mind than these evenings at Maer. I was also attached to and greatly revered my Uncle Jos: he was silent and reserved so as to be a rather awful man; but he sometimes talked openly with me. He was the very type of an upright man with the clearest judgement. I do not believe that any power on earth could have made him swerve an inch from what he considered the right course. I used to apply to him in my mind, the well-known ode of Horace, now forgotten by me, in which the words "nec vultus tyranni, &c.," come in.*

* CD is here referring to the third line from Horace, book 3, ode 3: 'Justum et tenacem'. Byron famously translated the passage as 'The man of firm and noble soul/ No factious clamors can control/ No threat'ning darkling brow/ Can swerve him from his just intent.'

Cambridge, 1828-1831

After having spent two sessions in Edinburgh, my father perceived or he heard from my sisters, that I did not like the thought of being a physician, so he proposed that I should become a clergyman. He was very properly vehement against my turning an idle sporting man, which then seemed my probable destination. I asked for some time to consider, as from what little I had heard and thought on the subject I had scruples about declaring my belief in all the dogmas of the Church of England; though otherwise I liked the thought of being a country clergyman. Accordingly I read with care Pearson on the Creeds and a few other books on divinity; and as I did not then in the least doubt the strict and literal truth of every word in the Bible, I soon persuaded myself that our Creed must be fully accepted. It never struck me how illogical it was to say that I believed in what I could not understand and what is in fact unintelligible. I might have said with entire truth that I had no wish to dispute any dogma; but I never was such a fool as to feel and say 'credo quia incredibile'.*

Considering how fiercely I have been attacked by the orthodox it seems ludicrous that I once intended to be a clergyman. Nor was this intention and my father's wish ever formally given up, but died a natural death when on leaving Cambridge I joined the *Beagle* as Naturalist. If the phrenologists are to be trusted, I was well fitted in one respect to be

* Editor's translation: 'I believe because it is unbelievable.' CD is using this phrase from the Latin Church Father Tertullian, without naming his source; he once again slightly misremembers his quotation, because the correct Latin is 'credo quia incredibilis'.

a clergyman. A few years ago the secretaries of a German psychological society asked me earnestly by letter for a photograph of myself, and some time afterwards I received the proceedings of one of the meetings in which it seemed that the shape of my head had been the subject of a public discussion, and one of the speakers declared that I had the bump of Reverence developed enough for ten Priests.

As it was decided that I should be a clergyman, it was necessary that I should go to one of the English universities and take a degree; but as I had never opened a classical book since leaving school, I found to my dismay that in the two intervening years I had actually forgotten, incredible as it may appear, almost everything which I had learnt even to some few of the Greek letters. I did not therefore proceed to Cambridge at the usual time in October, but worked with a private tutor in Shrewsbury and went to Cambridge after the Christmas vacation, early in 1828. I soon recovered my school standard of knowledge, and could translate easy Greek books, such as Homer and the Greek Testament with moderate facility.

During the three years which I spent at Cambridge my time was wasted, as far as the academical studies were concerned, as completely as at Edinburgh and at school. I attempted mathematics, and even went during the summer of 1828 with a private tutor (a very dull man) to Barmouth, but I got on very slowly. The work was repugnant to me, chiefly from my not being able to see any meaning in the early steps in algebra. This impatience was very foolish, and in after years I have deeply regretted that I did not proceed far enough at least to understand something of the great leading principles of mathematics; for men thus endowed seem to have an extra sense. But I do not believe that I should ever have succeeded beyond a very low grade. With respect to Classics I did nothing except attend a few compulsory college lectures, and the attendance was almost nominal. In my second year I had to work for a month or two to pass the Little Go, which I did easily. Again in my last year I worked with some earnestness for my final degree of B.A., and brushed up my Classics together with a little Algebra and Euclid, which latter gave me much pleasure, as it did whilst at school. In order to pass the B.A. examination, it was, also, necessary to get up Paley's *Evidences of Christianity*, and his *Moral Philosophy*. This was done in a thorough manner, and I am convinced that I could have written out the whole of the *Evidences* with perfect correctness, but not of course in the clear language of Paley. The logic of this book and as I may add of his *Natural Theology* gave me as much delight as did Euclid. The careful study of these works, without attempting to learn any part by rote, was the only part of the Academical Course which, as I then felt and as I still believe, was of the least use to me in the education of my mind. I did not at that time trouble myself about Paley's premises; and taking these on trust I was charmed and convinced by the long line of argumentation. By answering well the examination questions in Paley, by doing Euclid well, and by not failing miserably in Classics, I gained a good place among the *hoi polloi*, or crowd of men who do not go in for honours. Oddly enough I cannot remember how high I stood, and my memory fluctuates between the fifth, tenth, or twelfth name on the list.

Public lectures on several branches were given in the University, attendance being quite voluntary; but I was so sickened with lectures at Edinburgh that I did not even attend Sedgwick's eloquent and interesting lectures. Had I done so I should probably have become a geologist earlier than I did. I attended, however, Henslow's lectures on Botany, and liked them much for their extreme clearness, and the admirable illustrations; but I did not study botany. Henslow used to take his pupils, including several of the older members of the University, field excursions, on foot, or in coaches to distant places, or in a barge down the river, and lectured on the rarer plants or animals which were observed. These excursions were delightful.

Although as we shall presently see there were some redeeming features in my life at Cambridge, my time was sadly wasted there and worse than wasted. From my passion for shooting and for hunting and when this failed for riding across country I got into a sporting set, including some dissipated low-minded young men. We used often to dine together in the evening, though these dinners often included men of a higher stamp, and we sometimes drank too much, with jolly singing and playing at cards afterwards. I know that I ought to feel ashamed of days and evenings thus spent, but as some of my friends were very pleasant and we were all in the highest spirits, I cannot help looking back to these times with much pleasure.

But I am glad to think that I had many other friends of a widely different nature. I was very intimate with Whitley, who was afterwards Senior Wrangler, and we used continually to take long walks together. He inoculated me with a taste for pictures and good engravings, of which I bought some. I frequently went to the Fitzwilliam Gallery, and my taste must have been fairly good, for I certainly admired the best pictures, which I discussed with the old curator. I read also with much interest Sir J. Reynolds' book. This taste, though not natural to me, lasted for several years and many of the pictures in the National Gallery in London gave me much pleasure; that of Sebastian del Piombo exciting in me a sense of sublimity.

I also got into a musical set, I believe by means of my warm-hearted friend Herbert, who took a high wrangler's degree. From associating with these men and hearing them play, I acquired a strong taste for music, and used very often to time my walks so as to hear on week days the anthem in King's College Chapel. This gave me intense pleasure, so that my backbone would sometimes shiver. I am sure that there was no affectation or mere imitation in this taste, for I used generally to go by myself to King's College, and I sometimes hired the chorister boys to sing in my rooms. Nevertheless I am so utterly destitute of an ear, that I cannot perceive a discord, or keep time and hum a tune correctly; and it is a mystery how I could possibly have derived pleasure from music.

My musical friends soon perceived my state, and sometimes amused themselves by making me pass an examination, which consisted in ascertaining how many tunes I could recognise, when they were played rather more quickly or slowly than usual. 'God save the King' when thus played was a sore puzzle. There was another man with almost as bad an ear as I had, and strange to say he played a little on the flute. Once I had the triumph of beating him in one of our musical examinations.

But no pursuit at Cambridge was followed with nearly so much eagerness or gave me so much pleasure as collecting beetles. It was the mere passion for collecting, for I did not dissect them and rarely compared their external characters with published descriptions, but got them named anyhow. I will give a proof of my zeal: one day, on tearing off some old bark, I saw two rare beetles and seized one in each hand; then I saw a third and new kind, which I could not bear to lose, so that I popped the one which I held in my right hand into my mouth. Alas it ejected some intensely acrid fluid, which burnt my tongue so that I was forced to spit the beetle out, which was lost, as well as the third one. I was very successful in collecting and invented two new methods; I employed a labourer to scrape during the winter, moss off old trees and place it in a large bag, and likewise to collect the rubbish at the bottom of the barges in which reeds are brought from the fens, and thus I got some very rare species. No poet ever felt more delight at seeing his first poem published than I did at seeing in Stephen's *Illustrations of British Insects* the magic words, "captured by C. Darwin, Esq." I was introduced to entomology by my second cousin, W. Darwin Fox, a clever and most pleasant man, who was then at Christ's College and with whom I became extremely intimate. Afterwards I became well acquainted with and went out collecting, with Albert Way of Trinity, who in after years became a well-known archaeologist; also with H. Thompson, of the same College, afterwards a leading agriculturist, chairman of a great Railway, and Member of Parliament. It seems therefore that a taste for collecting beetles is some indication of future success in life!

I am surprised what an indelible impression many of the beetles which I caught at Cambridge have left on my mind. I can remember the exact appearance of certain posts, old trees and banks where I made a good capture. The pretty *Panagæus crux-major* was a treasure in those days, and here at Down I saw a beetle running across a walk, and on picking it up instantly perceived that it differed slightly from *P. crux-major*, and it turned out to be *P. quadripunctatus*, which is only a variety or closely allied species, differing from it very slightly in outline. I had never seen in those old days Licinus alive, which to an uneducated eye hardly differs from many other black Carabidous beetles; but my sons found here a specimen and I instantly recognised that it was new to me; yet I had not looked at a British beetle for the last twenty years.

I have not as yet mentioned a circumstance which influenced my whole career more than any other. This was my friendship with Prof. Henslow. Before coming up to Cambridge, I had heard of him from my brother as a man who knew every branch of science, and I was accordingly prepared to reverence him. He kept open house once every week, where all undergraduates and several older members of the University, who were attached to science, used to meet in the evening. I soon got, through Fox, an invitation, and went there regularly. Before long I became well acquainted with Henslow, and during the latter half of my time at Cambridge took long walks with him on most days; so that I was called by some of the dons "the man who walks with Henslow"; and in the evening I was very often asked to join his family dinner. His knowledge was great in botany, entomology, chemistry, mineralogy, and geology. His strongest taste was to draw conclusions from long-continued minute observations. His judgement was excellent, and his whole mind well-balanced; but I do not suppose that anyone would say that he possessed much original genius. He was deeply religious, and so orthodox, that he told me one day, he should be grieved if a single word of the 39 Articles were altered. His moral qualities were in every way admirable. He was free from every tinge of vanity or other petty feeling; and I never saw a man who thought so little about himself or his own concerns. His temper was imperturbably good, with the most winning and courteous manners; yet, as I have seen, he could be roused by any bad action to the warmest indignation and prompt action.

¶ I once saw in his company in the streets of Cambridge almost as horrid a scene, as could have been witnessed during the French Revolution. Two body-snatchers had been arrested and whilst being taken to prison had been torn from the constable by a crowd of the roughest men, who dragged them by their legs along the muddy and stony road. They were covered from head to foot with mud and their faces were bleeding either from having been kicked or from the stones; they looked like corpses, but the crowd was so dense that I got only a few momentary glimpses of the wretched creatures. Never in my life have I seen such wrath painted on a man's face, as was shown by Henslow at this horrid scene. He tried repeatedly to penetrate the mob; but it was simply impossible. He then rushed away to the mayor, telling me not to follow him, to get more policemen. I forget the issue, except that the two were got into the prison before being killed. Ω

Henslow's benevolence was unbounded, as he proved by his many excellent schemes for his poor parishioners, when in after years he held the living of Hitcham. My intimacy with such a man ought to have been and I hope was an inestimable benefit. I cannot resist mentioning a trifling incident, which showed his kind consideration. Whilst examining some pollen-grains on a damp surface I saw the tubes exserted, and instantly rushed off to communicate my surprising discovery to him. Now I do not suppose any other Professor of Botany could have helped laughing at my coming in such a hurry to make such a communication. But he agreed how interesting the phenomenon was, and explained its meaning, but made me clearly understand how well it was known; so I left him not in the least mortified, but well pleased at having discovered for myself so remarkable a fact, but determined not to be in such a hurry again to communicate my discoveries.

Dr. Whewell was one of the older and distinguished men who sometimes visited Henslow, and on several occasions I walked home with him at night. Next to Sir J. Mackintosh he was the best converser on grave subjects to whom I ever listened. Leonard Jenyns, (grandson of the famous Soames Jenyns), who afterwards published some good essays in Natural History, often staid with Henslow, who was his brother-in-law. At first I disliked him from his somewhat grim and sarcastic expression; and it is not often that a first impression is lost; but I was completely mistaken and found him very kindhearted, pleasant and with a good stock of humour. I visited him at his parsonage on the borders of the Fens (Swaffham Bulbeck) and had many a good walk and talk with him about Natural History. I became also acquainted with several other men older than me, who did not care much about science, but were friends of Henslow. One was a Scotchman, brother of Sir Alexander Ramsay, and tutor of Jesus College; he was a delightful man, but did not live for many years. Another was Mr Dawes, afterwards Dean of Hereford and famous for his success in the education of the poor. These men and others of the same standing, together with Henslow, used sometimes to take distant excursions into the country, which I was allowed to join and they were most agreeable.

Looking back, I infer that there must have been something in me a little superior to the common run of youths, otherwise the above-mentioned men, so much older than me and higher in academical position, would never have allowed me to associate with them. Certainly I was not aware of any such superiority, and I remember one of my sporting friends, Turner, who saw me at work on my beetles, saying that I should some day be a Fellow of the Royal Society, and the notion seemed to me preposterous.

During my last year at Cambridge I read with care and profound interest Humboldt's *Personal Narrative*. This work and Sir J. Herschel's *Introduction to the Study of Natural Philosophy* stirred up in me a burning zeal to add even the most humble contribution to the noble structure of Natural Science. No one or a dozen other books influenced me nearly so much as these two. I copied out from Humboldt long passages about Teneriffe, and read them aloud on one of the above-mentioned excursions, to (I think) Henslow, Ramsay and Dawes; for on a previous occasion I had talked about the glories of Teneriffe, and some of the party declared they would endeavour to go there; but I think that they were only half in earnest. I was, however, quite in earnest, and got an introduction to a merchant in London to enquire about ships; but the scheme was of course knocked on the head by the voyage of the *Beagle*.

My summer vacations were given up to collecting beetles, to some reading and short tours. In the autumn my whole time was devoted to shooting, chiefly at Woodhouse and Maer, and sometimes with young Eyton of Eyton. Upon the whole the three years which I spent at Cambridge were the most joyful in my happy life; for I was then in excellent health, and almost always in high spirits.

As I had at first come up to Cambridge at Christmas, I was forced to keep two terms after passing my final examination, at the commencement of 1831; and Henslow then persuaded me to begin the study of geology. Therefore on my return to Shropshire I examined sections and coloured a map of parts round Shrewsbury. Professor Sedgwick intended to visit

N. Wales in the beginning of August to pursue his famous geological investigation amongst the older rocks, and Henslow asked him to allow me to accompany him. Accordingly he came and slept at my Father's house.

A short conversation with him during this evening produced a strong impression on my mind. Whilst examining an old gravel-pit near Shrewsbury a labourer told me that he had found in it a large worn tropical Volute shell, such as may be seen on the chimney-pieces of cottages; and as he would not sell the shell I was convinced that he had really found it in the pit. I told Sedgwick of the fact, and he at once said (no doubt truly) that it must have been thrown away by someone into the pit; but then added, if really embedded there it would be the greatest misfortune to geology, as it would overthrow all that we know about the superficial deposits of the midland counties. These gravel-beds belonged in fact to the glacial period, and in after years I found in them broken arctic shells. But I was then utterly astonished at Sedgwick not being delighted at so wonderful a fact as a tropical shell being found near the surface in the middle of England. Nothing before had ever made me thoroughly realise, though I had read various scientific books, that science consists in grouping facts so that general laws or conclusions may be drawn from them.

Next morning we started for Llangollen, Conway, Bangor, and Capel Curig. This tour was of decided use in teaching me a little how to make out the geology of a country. Sedgwick often sent me on a line parallel to his, telling me to bring back specimens of the rocks and to mark the stratification on a map. I have little doubt that he did this for my good, as I was too ignorant to have aided him. On this tour I had a striking instance how easy it is to overlook phenomena, however conspicuous, before they have been observed by anyone. We spent many hours in Cwm Idwal, examining all the rocks with extreme care, as Sedgwick was anxious to find fossils in them; but neither of us saw a trace of the wonderful glacial phenomena all around us; we did not notice the plainly scored rocks, the perched boulders, the lateral and terminal moraines. Yet these phenomena are so conspicuous that, as I declared in a paper published many years afterwards in the *Philosophical Magazine*, a house burnt down by fire did not tell its story more plainly than did this valley. If it had still been filled by a glacier, the phenomena would have been less distinct than they now are.

At Capel Curig I left Sedgwick and went in a straight line by compass and map across the mountains to Barmouth, never following any track unless it coincided with my course. I thus came on some strange wild places and enjoyed much this manner of travelling. I visited Barmouth to see some Cambridge friends who were reading there, and thence returned to Shrewsbury and to Maer for shooting; for at that time I should have thought myself mad to give up the first days of partridge-shooting for geology or any other science.

Voyage of the *Beagle*:
from Dec. 27, 1831 to Oct. 2, 1836

On returning home from my short geological tour in N. Wales, I found a letter from Henslow, informing me that Captain FitzRoy was willing to give up part of his own cabin to any young man who would volunteer to go with him without pay as naturalist to the Voyage of the *Beagle*. I have given as I believe in my M.S. Journal an account of all the circumstances which then occurred; I will here only say that I was instantly eager to accept the offer, but my Father strongly objected, adding the words fortunate for me, – "If you can find any man of common sense, who advises you to go, I will give my consent." So I wrote that evening and refused the offer. On the next morning I went to Maer to be ready for September 1st, and whilst out shooting, my uncle sent for me, offering to drive me over to Shrewsbury and talk with my father. As my uncle thought it would be wise in me to accept the offer, and as my father always maintained that he was one of the most sensible men in the world, he at once consented in the kindest manner. I had been rather extravagant at Cambridge and to console my father said "that I should be deuced clever to spend more than my allowance whilst on board the *Beagle*"; but he answered with a smile, "But they all tell me you are very clever."

Next day I started for Cambridge to see Henslow, and thence to London to see FitzRoy, and all was soon arranged. Afterwards on becoming very intimate with FitzRoy, I heard that I had run a very narrow risk of being rejected, on account of the shape of my nose! He was an ardent disciple of Lavater and was convinced that he could judge a man's character by the outline of his features; and he doubted whether anyone with my nose could possess sufficient energy and determination for the voyage. But I think he was afterwards well-satisfied that my nose had spoken falsely.

¶ FitzRoy's character was a singular one, with many very noble features: he was devoted to his duty, generous to a fault, bold, determined, indomitably energetic, and an ardent friend to all under his sway. He would undertake any sort of trouble to assist those whom he thought deserved assistance. He was a handsome man, strikingly like a gentleman, with highly courteous manners, which resembled those of his maternal uncle, the famous Lord Castlereagh, as I was told by the Minister at Rio. Nevertheless he must have inherited much in his appearance from Charles II, for Dr. Wallich gave me a collection of photographs which he had made, and I was struck with the resemblance of one to FitzRoy; on looking at the name, I found it Ch. E. Sobieski Stuart, Count d'Albanie, illegitimate descendant of the same monarch.

FitzRoy's temper was a most unfortunate one. This was shown not only by passion but by fits of long-continued moroseness against those who had offended him. His temper was usually worst in the early morning, and with his eagle eye he could generally detect something amiss about the ship, and was then unsparing in his blame. The junior officers when they relieved each other in the forenoon used to ask "whether much hot coffee had been served out this morning, – " which meant how was the Captain's temper? He was also somewhat suspicious and occasionally in very low spirits, on one occasion bordering on insanity. He seemed to me often to fail in sound judgment or common sense. He was* extremely kind to me, but was a man very difficult to live with on the intimate terms which necessarily followed from our messing by ourselves in the same cabin. We had several quarrels; for when out of temper he was utterly unreasonable. For instance, early in the voyage at Bahia in Brazil he defended and praised slavery, which I abominated, and told me that he had just visited a great slave-owner, who had called up many of his slaves and asked them whether they were happy, and whether they wished to be free, and all answered "No." I

* 'very' deleted.

then asked him, perhaps with a sneer, whether he thought that the answers of slaves in the presence of their master was worth anything. This made him excessively angry, and he said that as I doubted his word, we could not live any longer together. I thought that I should have been compelled to leave the ship; but as soon as the news spread, which it did quickly, as the captain sent for his first lieutenant to assuage his anger by abusing me, I was deeply gratified by receiving an invitation from all the gun-room officers to mess with them. But after a few hours FitzRoy showed his usual magnanimity by sending an officer to me with an apology and a request that I would continue to live with him. I remember another instance of his candour. At Plymouth before we sailed, he was extremely angry with a dealer in crockery who refused to exchange some article purchased in his shop: the Captain asked the man the price of a very expensive set of china and said "I should have purchased this if you had not been so disobliging." As I knew that the cabin was amply stocked with crockery, I doubted whether he had any such intention; and I must have shown my doubts in my face, for I said not a word. After leaving the shop he looked at me, saying You do not believe what I have said, and I was forced to own that it was so. He was silent for a few minutes and then said You are right, and I acted wrongly in my anger at the blackguard.

At Conception in Chile, poor FitzRoy was sadly overworked and in very low spirits; he complained bitterly to me that he must give a great party to all the inhabitants of the place. I remonstrated and said that I could see no such necessity on his part under the circumstances. He then burst out into a fury, declaring that I was the sort of man who would receive any favours and make no return. I got up and left the cabin without saying a word, and returned to Conception where I was then lodging. After a few days I came back to the ship and was received by the Captain as cordially as ever, for the storm had by that time quite blown over. The first Lieutenant, however, said to me: "Confound you, philosopher, I wish you would not quarrel with the skipper; the day you left the ship I was dead-tired (the ship was refitting) and he kept me walking the deck till midnight abusing you all the time." The difficulty of living on good terms with a Captain of a Man-of-War is much increased by its being almost mutinous to answer him as one would answer anyone else; and by the awe in which he is held – or was held in my time, by all on board. I remember hearing a curious instance of this in the case of the purser of the *Adventure*, – the ship which sailed with the *Beagle* during the first voyage. The Purser was in a store in Rio de Janeiro, purchasing rum for the ship's company, and a little gentleman in plain clothes walked in. The Purser said to him, "Now Sir, be so kind as to taste this rum, and give me your opinion of it." The gentleman did as he was asked, and soon left the store. The store-keeper then asked the Purser, whether he knew that he had been speaking to the Captain of a Line of Battleships which had just come into the harbour. The poor Purser was struck dumb with horror; he let the glass of spirit drop from his hand onto the floor, and immediately went on board, and no persuasion, as an officer on the *Adventure* assured me, could make him go on shore again for fear of meeting the Captain after his dreadful act of familiarity.

I saw FitzRoy only occasionally after our return home, for I was always afraid of unintentionally offending him, and did so once, almost beyond mutual reconciliation. He was afterwards very indignant with me for having published so unorthodox a book (for he became very religious) as the *Origin of Species*. Towards the close of his life he was as I fear, much impoverished, and this was largely due to his generosity. Anyhow after his death a subscription was raised to pay his debts. His end was a melancholy one, namely suicide, exactly like that of his uncle Ld. Castlereagh, whom he resembled closely in manner and appearance.

His character was in several respects one of the most noble which I have ever known, though tarnished by grave blemishes. Ω

The voyage of the *Beagle* has been by far the most important event in my life and has determined my whole career; yet it depended on so small a circumstance as my uncle offering to drive me 30 miles to Shrewsbury, which few uncles would have done, and on such a trifle as the shape of my nose. I have always felt that I owe to the voyage the first real training or education of my mind. I was led to attend closely to several branches of natural history, and thus my powers of observation were improved, though they were already fairly developed.

The investigation of the geology of all the places visited was far more important, as reasoning here comes into play. On first examining a new district nothing can appear more hopeless than the chaos of rocks; but by recording the stratification and nature of the rocks and fossils at many points, always reasoning and predicting what will be found elsewhere, light soon begins to dawn on the district, and the structure of the whole becomes more or less intelligible. I had brought with me the first volume of Lyell's *Principles of Geology*, which I studied attentively; and this book was of the highest service to me in many ways. The very first place which I examined, namely St. Jago in the Cape Verde islands, showed me clearly the wonderful superiority of Lyell's manner of treating geology, compared with that of any other author, whose works I had with me or ever afterwards read.

Another of my occupations was collecting animals of all classes, briefly describing and roughly dissecting many of the marine ones; but from not being able to draw and from not having sufficient anatomical knowledge a great pile of MS. which I made during the voyage has proved almost useless. I thus lost much time, with the exception of that spent in acquiring some knowledge of the Crustaceans, as this was of service when in after years I undertook a monograph of the Cirripedia.

During some part of the day I wrote my Journal, and took much pains in describing carefully and vividly all that I had seen; and this was good practice. My Journal served, also, in part as letters to my home, and portions were sent to England, whenever there was an opportunity.

The above various special studies were, however, of no importance compared with the habit of energetic industry and of concentrated attention to whatever I was engaged in, which I then acquired. Everything about which I thought or read was made to bear directly on what I had seen and was likely to see; and this habit of mind was continued during the five years of the voyage. I feel sure that it was this training which has enabled me to do whatever I have done in science.

Looking backwards, I can now perceive how my love for science gradually preponderated over every other taste. During the first two years my old passion for shooting survived in nearly full force, and I shot myself all the birds and animals for my collection; but gradually I gave up my gun more and more, and finally altogether to my servant, as shooting interfered with my work, more especially with making out the geological structure of a country. I discovered, though unconsciously and insensibly, that the pleasure of observing and reasoning was a much higher one than that of skill and sport. ¶ The primeval instincts of the barbarian slowly yielded to the acquired tastes of the civilized man. Ω That my mind became developed through the pursuits during the voyage, is rendered probable by a remark made by my father, who was the most acute observer whom I ever saw, of a sceptical disposition, and far from being a believer in phrenology; for on first seeing me after the voyage, he turned round to my sisters and exclaimed, "Why, the shape of his head is quite altered."

To return to the voyage. On September 11th (1831) I paid a flying visit with FitzRoy to the *Beagle* at Plymouth. Thence to Shrewsbury to wish my father and sisters a long farewell. On Oct. 24th, I took up my residence at Plymouth, and remained there until December 27th when the *Beagle* finally left the shores of England for her circumnavigation of the world. We made two earlier attempts to sail, but were driven back each time by heavy gales. These two months at Plymouth were the most miserable which I ever spent, though I exerted myself in various ways. I was out of spirits at the thought of leaving all my family and friends for so long a time, and the weather seemed to me inexpressibly gloomy. I was also troubled with palpitations and pain about the heart, and like many a young ignorant man, especially one with a smattering of medical knowledge, was convinced that I had heart-disease. I did not consult any doctor, as I fully expected to hear the verdict that I was not fit for the voyage, and I was resolved to go at all hazards.

I need not here refer to the events of the voyage – where we went and what we did – as I have given a sufficiently full account in my published Journal. The glories of the vegetation of the Tropics rise before my mind at the present time more vividly than anything else. Though the sense of sublimity, which the great deserts of Patagonia and the forest-clad mountains of Tierra del Fuego excited in me, has left an indelible impression on my mind. The sight of a naked savage in his native land is an event which can never be forgotten. Many of my excursions on horseback through wild countries, or in the boats, some of which lasted several weeks, were deeply interesting; their discomfort and some degree of danger were at that time hardly a drawback and none at all afterwards. I also reflect with high satisfaction on some of my scientific work, such as solving the problem of coral-islands, and making out the geological structure of certain islands, for instance, St. Helena. ¶ Nor must I pass over the discovery of the singular relations of the animals and plants inhabiting the several islands of the Galapagos archipelago, and of all of them to the inhabitants of South America. Ω

As far as I can judge of myself I worked to the utmost during the voyage from the mere pleasure of investigation, and from my strong desire to add a few facts to the great mass of facts in natural science. But I was also ambitious to take a fair place among scientific men, – whether more ambitious or less so than most of my fellow-workers I can form no opinion.

The geology of St. Jago is very striking yet simple: a stream of lava formerly flowed over the bed of the sea, formed of triturated recent shells and corals, which it has baked into a hard white rock. Since then the whole island has been upheaved. But the line of white rock revealed to me a new and important fact, namely that there had been afterwards subsidence round the craters, which had since been in action, and had poured forth lava. It then first dawned on me that I might perhaps write a book on the geology of the various countries visited, and this made me thrill with delight. That was a memorable hour to me, and how distinctly I can call to mind the low cliff of lava beneath which I rested, with the sun glaring hot, a few strange desert plants growing near, and with living corals in the tidal pools at my feet. Later in the voyage FitzRoy asked to read some of my Journal, and declared it would be worth publishing; so here was a second book in prospect!

Towards the close of our voyage I received a letter whilst at Ascension, in which my sisters told me that Sedgwick had called on my father and said that I should take a place among the leading scientific men. I could not at the time understand how he could have learnt anything of my proceedings, but I heard (I believe afterwards) that Henslow had read some of the letters which I wrote to him before the Philosophical

Soc. of Cambridge and had printed them for private distribution. My
collection of fossil bones, which had been sent to Henslow, also excited
considerable attention amongst palæontologists. After reading this letter
I clambered over the mountains of Ascension with a bounding step and
made the volcanic rocks resound under my geological hammer! All this
shows how ambitious I was; but I think that I can say with truth that in
after years, though I cared in the highest degree for the approbation of
such men as Lyell and Hooker, who were my friends, I did not care
much about the general public. I do not mean to say that a favourable
review or a large sale of my books did not please me greatly; but the
pleasure was a fleeting one, and I am sure that I have never turned one
inch out of my course to gain fame.

From my return to England
Oct. 2 1836 to my marriage Jan. 29 1839

These two years and three months were the most active ones which I ever spent, though I was occasionally unwell and so lost some time. After going backwards and forwards several times between Shrewsbury, Maer, Cambridge and London, I settled in lodgings at Cambridge on December 13th, where all my collections were under the care of Henslow. I stayed here three months and got my minerals and rocks examined by the aid of Prof. Miller. I began preparing my Journal of travels, which was not hard work, as my MS Journal had been written with care, and my chief labour was making an abstract of my more interesting scientific results. I sent also, at the request of Lyell, a short account of my observations on the elevation of the coast of Chile to the Geological Society.

On March 7th, 1837, I took lodgings in Great Marlborough Street in London and remained there for nearly two years until I was married. During these two years I finished my Journal, read several papers before the Geological Society, began preparing the MS for my *Geological Observations* and arranged for the publication of the *Zoology of the Voyage of the Beagle*. In July I opened my first note-book for facts in relation to the *Origin of Species*, about which I had long reflected, and never ceased working on for the next twenty years.

During these two years I also went a little into society, and acted as one of the hon. secretaries of the Geological Society. I saw a great deal of Lyell. One of his chief characteristics was his sympathy with the work of others; and I was as much astonished as delighted at the interest which he showed when on my return to England I explained to him my views on coral reefs. This encouraged me greatly, and his advice and example had much influence on me. During this time I saw also a good deal of Robert Brown "facile princeps botanicorum."* I used often to call and sit with him during his breakfast on Sunday mornings, and he poured forth a rich treasure of curious observations and acute remarks, but they almost always related to minute points, and he never with me discussed large and general questions in science.

During these two years I took several short excursions as a relaxation, and one longer one to the parallel roads of Glen Roy, an account of which was published in the *Philosophical Transactions*. This paper was a great failure, and I am ashamed of it. Having been deeply impressed with what I had seen of the elevation of the land in S. America, I attributed the parallel lines to the action of the sea; but I had to give up this view when Agassiz propounded his glacier-lake theory. Because no other explanation was possible under our then state of knowledge, I argued in favour of sea-action; and my error has been a good lesson to me never to trust in science to the principle of exclusion.

As I was not able to work all day at science I read a good deal during these two years on various subjects, including some metaphysical books, but I was not at all well fitted for such studies. About this time I took much delight in Wordsworth's and Coleridge's poetry, and can boast that I read the *Excursion* twice through. Formerly Milton's *Paradise Lost* had been my chief favourite, and in my excursions during the voyage of the *Beagle*, when I could take only a single small volume, I always chose Milton.

* CD is Latinizing 'Unquestionably the most eminent of botanists'.

Religious Belief

During these two years I was led to think much about religion. Whilst on board the *Beagle* I was quite orthodox, and I remember being heartily laughed at by several of the officers (though themselves orthodox) for quoting the Bible as an unanswerable authority on some point of morality. I suppose it was the novelty of the argument that amused them. But I had gradually come, by this time, to see that the Old Testament from its manifestly false history of the world, with the Tower of Babel, the rainbow as a sign, etc., etc., and from its attributing to God the feelings of a revengeful tyrant, was no more to be trusted than the sacred books of the Hindoos, or the beliefs of any barbarian. The question then continually rose before my mind and would not be banished, – is it credible that if God were now to make a revelation to the Hindoos, would he permit it to be connected with the belief in Vishnu, Siva, &c., as Christianity is connected with the Old Testament. This appeared to me utterly incredible.

By further reflecting that the clearest evidence would be requisite to make any sane man believe in the miracles by which Christianity is supported, – that the more we know of the fixed laws of nature the more incredible do miracles become, – that the men at that time were ignorant and credulous to a degree almost incomprehensible by us, – that the Gospels cannot be proved to have been written simultaneously with the events, – that they differ in many important details, far too important as it seemed to me to be admitted as the usual inaccuracies of eyewitnesses; — by such reflections as these, which I give not as having the least novelty or value, but as they influenced me, I gradually came to disbelieve in Christianity as a divine revelation. The fact that many false religions have spread over large portions of the earth like wild-fire had some weight with me. Beautiful as is the morality of the New Testament, it can hardly be denied that its perfection depends in part on the interpretation which we now put on metaphors and allegories.

But I was very unwilling to give up my belief; – I feel sure of this for I can well remember often and often inventing day-dreams of old letters between distinguished Romans and manuscripts being discovered at Pompeii or elsewhere which confirmed in the most striking manner all that was written in the Gospels. But I found it more and more difficult, with free scope given to my imagination, to invent evidence which would suffice to convince me. Thus disbelief crept over me at a very slow rate, but was at last complete. The rate was so slow that I felt no distress, and have never since doubted even for a single second that my conclusion was correct. I can indeed hardly see how anyone ought to wish Christianity to be true; for if so the plain language of the text seems to show that the men who do not believe, and this would include my Father, Brother and almost all my best friends, will be everlastingly punished.

And this is a damnable doctrine.

Although I did not think much about the existence of a personal God until a considerably later period of my life, I will here give the vague conclusions to which I have been driven. The old argument of design in nature, as given by Paley, which formerly seemed to me so conclusive, fails, now that the law of natural selection has been discovered. We can no longer argue that, for instance, the beautiful hinge of a bivalve shell must have been made by an intelligent being, like the hinge of a door by man. There seems to be no more design in the variability of organic beings and in the action of natural selection, than in the course which the wind blows. Everything in nature is the result of fixed laws. But I have discussed this subject at the end of my book on the *Variation of Domestic Animals and Plants*, and the argument there given has never, as far as I can see, been answered.

But passing over the endless beautiful adaptations which we everywhere meet with, it may be asked how can the generally beneficent arrangement of the world be accounted for? Some writers indeed are so much impressed with the amount of suffering in the world, that they doubt if we look to all sentient beings, whether there is more of misery or of happiness;—whether the world as a whole is a good or a bad one. According to my judgment happiness decidedly prevails, though this would be very difficult to prove. If the truth of this conclusion be granted, it harmonises well with the effects which we might expect from natural selection. If all the individuals of any species were habitually to suffer to an extreme degree they would neglect to propagate their kind; but we have no reason to believe that this has ever or at least often occurred. Some other considerations, moreover, lead to the belief that all sentient beings have been formed so as to enjoy, as a general rule, happiness. Every one who believes, as I do, that all the corporeal and mental organs (excepting those which are neither advantageous or disadvantageous to the possessor) of all beings have been developed through natural selection, or the survival of the fittest, together with use or habit, will admit that these organs have been formed so that their possessors may compete successfully with other beings, and thus increase in number. Now an animal may be led to pursue that course of action which is the most beneficial to the species by suffering, such as pain, hunger, thirst, and fear, – or by pleasure, as in eating and drinking and in the propagation of the species, &c. or by both means combined, as in the search for food. But pain or suffering of any kind, if long continued, causes depression and lessens the power of action; yet is well adapted to make a creature guard itself against any great or sudden evil. Pleasurable sensations, on the other hand, may be long continued without any depressing effect; on the contrary they stimulate the whole system to increased action. Hence it has come to pass that most or all sentient beings have been developed in such a manner through natural selection, that pleasurable sensations serve as their habitual guides. We see this in the pleasure from exertion, even occasionally from great exertion of the body or mind, – in the pleasure of our daily meals, and especially in the pleasure derived from sociability and from loving our families. The sum of such pleasures as these, which are habitual or frequently recurrent, give, as I can hardly doubt, to most sentient beings an excess of happiness over misery, although many occasionally suffer much. Such suffering, is quite compatible with the belief in Natural Selection, which is not perfect in its action, but tends only to render each species as successful as possible in the battle for life with other species, in wonderfully complex and changing circumstances.

That there is much suffering in the world no one disputes. Some have attempted to explain this in reference to man by imagining that it serves for his moral improvement. But the number of men in the world is as nothing compared with that of all other sentient beings, and these often suffer greatly without any moral improvement. A being so powerful and so full of knowledge as a God who could create the universe, is to our finite minds omnipotent and omniscient, and it revolts our understanding to suppose that his benevolence is not unbounded, for what advantage can there be in the sufferings of millions of the lower animals throughout almost endless time? This very old argument from the existence of suffering against the existence of an intelligent first cause seems to me a strong one; whereas, as just remarked, the presence of much suffering agrees well with the view that all organic beings have been developed through variation and natural selection.

At the present day the most usual argument for the existence of an intelligent God is drawn from the deep inward conviction and feelings which are experienced by most persons. But it cannot be doubted that Hindoos, Mahomadans and others might argue in the same manner and with equal force in favour of the existence of one God, or of many Gods, or as with the Buddists of no God. There are also many barbarian tribes who cannot be said with any truth to believe in what we call God: they believe indeed in spirits or ghosts, and it can be explained, as Tyler and Herbert Spencer have shown, how such a belief would be likely to arise.

Formerly I was led by feelings such as those just referred to, (although I do not think that the religious sentiment was ever strongly developed in me), to the firm conviction of the existence of God, and of the immortality of the soul. In my Journal I wrote that whilst standing in the midst of the grandeur of a Brazilian forest, 'it is not possible to give an adequate idea of the higher feelings of wonder, admiration, and devotion which fill and elevate the mind.' I well remember my conviction that there is more in man than the mere breath of his body. But now the grandest scenes would not cause any such convictions and feelings to rise in my mind. It may be truly said that I am like a man who has become colour-blind, and the universal belief by men of the existence of redness makes my present loss of perception of not the least value as evidence. This argument would be a valid one if all men of all races had the same inward conviction of the existence of one God; but we know that this is very far from being the case. Therefore I cannot see that such inward convictions and feelings are of any weight as evidence of what really exists. The state of mind which grand scenes formerly excited in me, and which was intimately connected with a belief in God, did not essentially differ from that which is often called the sense of sublimity; and however difficult it may be to explain the genesis of this sense, it can hardly be advanced as an argument for the existence of God, any more than the powerful though vague and similar feelings excited by music.

¶ With respect to immortality, nothing shows me how strong and almost instinctive a belief it is, as the consideration of the view now held by most physicists, namely that the sun with all the planets will in time grow too cold for life, unless indeed some great body dashes into the sun and thus gives it fresh life. – Believing as I do that man in the distant future will be a far more perfect creature than he now is, it is an intolerable thought that he and all other sentient beings are doomed to complete annihilation after such long-continued slow progress. To those who fully admit the immortality of the human soul, the destruction of our world will not appear so dreadful. Ω

Another source of conviction in the existence of God, connected with the reason and not with the feelings, impresses me as having much more weight. This follows from the extreme difficulty or rather impossibility of conceiving this immense and wonderful universe, including man with his capacity of looking far backwards and far into futurity, as the result of blind chance or necessity. When thus reflecting I feel compelled to look to a First Cause having an intelligent mind in some degree analogous to that of man; and I deserve to be called a Theist.

¶ This conclusion was strong in my mind about the time, as far as I can remember, when I wrote the *Origin of Species*; and it is since that time that it has very gradually with many fluctuations become weaker.Ω

But then arises the doubt – can the mind of man, which has, as I fully believe, been developed from a mind as low as that possessed by the lowest animal, be trusted when it draws such grand conclusions? May not these be the result of the connection between cause and effect which strikes us as a necessary one, but probably depends merely on inherited experience? Nor must we overlook the probability of the constant inculcation in a belief in God on the minds of children producing so strong and perhaps an inherited effect on their brains not yet fully developed, that it would be as difficult for them to throw off their belief in God, as for a monkey to throw off its instinctive fear and hatred of a snake. I cannot pretend to throw the least light on such abstruse problems. The mystery of the beginning of all things is insoluble by us; and I for one must be content to remain an Agnostic.

A man who has no assured and ever present belief in the existence of a personal God or of a future existence with retribution and reward, can have for his rule of life, as far as I can see, only to follow those impulses and instincts which are the strongest or which seem to him the best ones. A dog acts in this manner, but he does so blindly. A man, on the other hand, looks forwards and backwards, and compares his various feelings, desires and recollections. He then finds, in accordance with the verdict of all the wisest men that the highest satisfaction is derived from following certain impulses, namely the social instincts. If he acts for the good of others, he will receive the approbation of his fellow men and gain the love of those with whom he lives; and this latter gain undoubtedly is the highest pleasure on this earth. By degrees it will become intolerable to him to obey his sensuous passions rather than his higher impulses, which when rendered habitual may be almost called instincts. His reason may occasionally tell him to act in opposition to the opinion of others, whose approbation he will then not receive; but he will still have the solid satisfaction of knowing that he has followed his inner-most guide or conscience.—As for myself I believe that I have acted rightly in steadily following and devoting my life to science. I feel no remorse from having committed any great sin, but have often and often regretted that I have not done more direct good to my fellow creatures. My sole and poor excuse is much ill-health and my mental constitution, which makes it extremely difficult for me to turn from one subject or occupation to another. I can imagine with high satisfaction giving up my whole time to philanthropy, but not a portion of it; though this would have been a far better line of conduct.

¶ Nothing is more remarkable than the spread of scepticism or rationalism during the latter half of my life. Before I was engaged to be married, my father advised me to conceal carefully my doubts, for he said that he had known extreme misery thus caused with married persons. Things went on pretty well until the wife or husband became out of health, and then some women suffered miserably by doubting about the salvation of their husbands, thus making them likewise to suffer. My father added that he had known during his whole long life only three women who were sceptics; and it should be remembered that he knew well a multitude of persons and possessed extraordinary power of winning confidence. When I asked him who the three women were, he had to own with respect to one of them, his sister-in-law Kitty Wedgwood, that he had no good evidence, only the vaguest hints, aided by the conviction that so clear-sighted a woman could not be a believer. At the present time, with my small acquaintance, I know (or have known) several married ladies, who believe very little more than their husbands. My father used to quote an unanswerable argument, by which an old lady, a Mrs Barlow, who suspected him of unorthodoxy, hoped to convert him:— "Doctor, I know that sugar is sweet in my mouth, and I know that my Redeemer liveth."* Ω

---

* CD adds a note: 'written in 1879 – copied out Ap 22 1881'.

From my marriage, Jan. 29 1839, and residence in Upper Gower St. to our leaving London and settling at Down, Sep. 14 1842

You all know well your Mother, and what a good Mother she has ever been to all of you. She has been my greatest blessing, and I can declare that in my whole life I have never heard her utter one word which I had rather have been unsaid. She has never failed in the kindest sympathy towards me, and has borne with the utmost patience my frequent complaints from ill-health and discomfort. I do not believe she has ever missed an opportunity of doing a kind action to anyone near her. I marvel at my good fortune that she, so infinitely my superior in every single moral quality, consented to be my wife. She has been my wise adviser and cheerful comforter throughout life, which without her would have been during a very long period a miserable one from ill-health. She has earned the love and admiration of every soul near her. (Mem: her beautiful letter to myself preserved, shortly after our marriage.)*

I have indeed been most happy in my family, and I must say to you my children that not one of you has ever given me one minute's anxiety, except on the score of health. There are, I suspect, very few fathers of five sons who could say this with entire truth. When you were very young it was my delight to play with you all, and I think with a sigh that such days can never return. From your earliest days to now that you are grown up, you have all, sons and daughters, ever been most pleasant, sympathetic and affectionate to us and to one another. When all or most of you are at home (as, thank Heavens happens pretty frequently) no party can be, according to my taste, more agreeable, and I wish for no

---

* CD added this memorandum in pencil.

other society. We have suffered only one very severe grief in the death of Annie at Malvern on April 24th, 1851, when she was just over ten years old. She was a most sweet and affectionate child, and I feel sure would have grown into a delightful woman. But I need say nothing here of her character, as I wrote a short sketch of it shortly after her death. Tears still sometimes come into my eyes, when I think of her sweet ways.

During the three years and eight months whilst we resided in London, I did less scientific work, though I worked as hard as I possibly could, than during any other equal length of time in my life. This was owing to frequently recurring unwellness and to one long and serious illness. The greater part of my time, when I could do anything, was devoted to my work on *Coral Reefs*, which I had begun before my marriage, and of which the last proof-sheet was corrected on May 6th, 1842. This book, though a small one, cost me twenty months of hard work, as I had to read every work on the islands of the Pacific and to consult many charts. It was thought highly of by scientific men, and the theory therein given is, I think, now well established.

No other work of mine was begun in so deductive a spirit as this; for the whole theory was thought out on the west coast of S. America before I had seen a true coral reef. I had therefore only to verify and extend my views by a careful examination of living reefs. But it should be observed that I had during the two previous years been incessantly attending to the effects on the shores of S. America of the intermittent elevation of the land, together with denudation and the deposition of sediment. This necessarily led me to reflect much on the effects of subsidence, and it was easy to replace in imagination the continued deposition of sediment by the upward growth of coral. To do this was to form my theory of the formation of barrier-reefs and atolls.

Besides my work on coral-reefs, during my residence in London, I read before the Geological Society papers on the Erratic Boulders of S. America, on Earthquakes, and on the Formation by the Agency of Earth-worms of Mould. I also continued to superintend the publication of the *Zoology of the Voyage of the Beagle*. Nor did I ever intermit collecting facts bearing on the origin of species; and I could sometimes do this when I could do nothing else from illness.

In the summer of 1842 I was stronger than I had been for some time and took a little tour by myself in N. Wales, for the sake of observing the effects of the old glaciers which formerly filled all the larger valleys. I published a short account of what I saw in the *Philosophical Magazine*. This excursion interested me greatly, and it was the last time I was ever strong enough to climb mountains or to take long walks, such as are necessary for geological work.

During the early part of our life in London, I was ¶* strong enough to go into general society, and saw a good deal of several scientific men and other more or less distinguished men. I will give my impressions with respect to some of them, though I have little to say worth saying.

I saw more of Lyell than of any other man both before and after my marriage. His mind was characterised, as it appeared to me, by clearness, caution, sound judgement and a good deal of originality. When I made any remark to him on Geology, he never rested until he saw the whole case clearly and often made me see it more clearly than I had done before. He would advance all possible objections to my suggestion, and even after these were exhausted would long remain dubious. A second characteristic was his hearty sympathy with the work of other scientific men.

On my return from the voyage of the *Beagle*, I explained to him my views on coral-reefs, which differed from his, and I was greatly surprised and encouraged by the vivid interest which he showed. On such occasions, while absorbed in thought, he would throw himself into the strangest attitudes, often resting his head on the seat of a chair, while standing up. His delight in science was ardent, and he felt the keenest interest in the future progress of mankind. He was very kind-hearted, and thoroughly liberal in his religious beliefs or rather disbeliefs; but he was a strong theist. His candour was highly remarkable. He exhibited this by becoming a convert to the Descent-theory, though he had gained much fame by opposing Lamarck's views, and this after he had grown old. He reminded me that I had many years before said to him, when discussing the opposition of the old school of geologists to his new

* CD writes here 'written 1881'.

views, "What a good thing it would be, if every scientific man was to die when sixty years old, as afterwards he would be sure to oppose all new doctrines." But he hoped that now he might be allowed to live. He had a strong sense of humour and often told amusing anecdotes. He was very fond of society, especially of eminent men, and of persons high in rank; and this over-estimation of a man's position in the world, seemed to me his chief foible. He used to discuss with Lady Lyell as a most serious question, whether or not they should accept some particular invitation. But as he would not dine out more than three times a week on account of the loss of time, he was justified in weighing his invitations with some care. He looked forward to going out oftener in the evening with advancing years, as to a great reward; but the good time never came, as his strength failed.

The science of Geology is enormously indebted to Lyell – more so, as I believe, than to any other man who ever lived. When I was starting on the voyage of the *Beagle*, the sagacious Henslow, who, like all other geologists believed at that time in successive cataclysms, advised me to get and study the first volume of the <u>Principles</u>, which had then just been published, but on no account to accept the views therein advocated. How differently would any one now speak of the <u>Principles</u>! I am proud to remember that the first place, namely St. Jago, in the Cape Verde Archipelago, which I geologised, convinced me of the infinite superiority of Lyell's views over those advocated in any other work known to me.

The powerful effects of Lyell's works could formerly be plainly seen in the different progress of the science in France and England. The present total oblivion of Elie de Beaumont's wild hypotheses, such as his *Craters of Elevation* and *Lines of Elevation* (which latter hypothesis I heard Sedgwick at the Geolog. Soc. lauding to the skies), may be largely attributed to Lyell.

All the leading geologists were more or less known by me, at the time when geology was advancing with triumphant steps. I liked most of them, with the exception of Buckland, who though very good-humoured and good-natured seemed to me a vulgar and almost coarse man. He was incited more by a craving for notoriety, which sometimes made him act like a buffoon, than by a love of science. He was not, however, selfish in his desire for notoriety; for Lyell, when a very young man, consulted him about communicating a poor paper to the Geol. Soc. which had been sent him by a stranger, and Buckland answered – "You had better do so, for it will be headed, 'Communicated by Charles Lyell', and thus your name will be brought before the public."

The services rendered to geology by Murchison by his classification of the older formations cannot be over-estimated; but he was very far from possessing a philosophical mind. He was very kind-hearted and would exert himself to the utmost to oblige anyone. The degree to which he valued rank was ludicrous, and he displayed this feeling and his vanity with the simplicity of a child. He related with the utmost glee to a large circle, including many mere acquaintances, in the rooms of the Geolog. Soc. how the Czar Nicholas, when in London, had patted him on the shoulder and had said, alluding to his geological work – "Mon ami, Russia is grateful to you," and then Murchison added rubbing his hands together, "The best of it was that Prince Albert heard it all." He announced one day to the Council of the Geolog. Soc. that his great work on the Silurian system was at last published; and he then looked at all who were present and said, "You will every one of you find your name in the Index," as if this was the height of glory.

I saw a good deal of Robert Brown, "facile Princeps Botanicorum," as he was called by Humboldt; and before I was married I used to go and sit with him almost every Sunday morning. He seemed to me to be chiefly remarkable for the minuteness of his observations and their perfect accuracy. He never propounded to me any large scientific views in biology. His knowledge was extraordinarily great, and much died with him, owing to his excessive fear of ever making a mistake. He poured out his knowledge to me in the most unreserved manner, yet was strangely jealous on some points. I called on him two or three times before the voyage of the *Beagle*, and on one occasion he asked me to look through a microscope and describe what I saw. This I did, and believe now that it was the marvellous currents of protoplasm in some vegetable cell. I then asked him what I had seen; but he answered me, who was then hardly more than a boy and on the point of leaving England for five years, "That is my little secret." I suppose that he was afraid that I might steal his discovery. Hooker told me that he was a complete miser, and knew himself to be a miser, about his dried plants; and he would not lend specimens to Hooker, who was describing the plants of Tierra del Fuego, although well knowing that he himself would never make any use of the collections from this country. On the other hand he was capable of the most generous actions. When old, much out of health and quite unfit for any exertion, he daily visited (as Hooker told me) an old man-servant, who lived at a distance and whom he supported, and read aloud to him. This is enough to make up for any degree of scientific penuriousness or jealousy. He was rather given to sneering at anyone who wrote about what he did not fully understand: I remember praising Whewell's *History of the Inductive Sciences* to him, and he answered, "Yes, I suppose that he has read the prefaces of very many books."

I often saw Owen, whilst living in London, and admired him greatly, but was never able to understand his character and never became intimate with him. After the publication of the *Origin of Species* he became my bitter enemy, not owing to any quarrel between us, but as far as I could judge out of jealousy at its success. Poor dear Falconer, who was a charming man, had a very bad opinion of him, being convinced that he was not only ambitious, very envious and arrogant, but untruthful and dishonest. His power of hatred was certainly unsurpassed. When in former days I used to defend Owen, Falconer often said, "You will find him out some day," and so it has proved.

At a somewhat later period I became very intimate with Hooker, who has been one of my best friends throughout life. He is a delightfully pleasant companion and most kind-hearted. One can see at once that he is honourable to the back-bone. His intellect is very acute, and he has great power of generalisation. He is the most untirable worker that I have ever seen, and will sit the whole day working with the microscope, and be in the evening as fresh and pleasant as ever. He is in all ways very impulsive and somewhat peppery in temper; but the clouds pass away almost immediately. He once sent me an almost savage letter from a cause which will appear ludicrously small to an outsider, viz. because I maintained for a time the silly notion that our coal-plants had lived in shallow water in the sea. His indignation was all the greater because he could not pretend that he should ever have suspected that the Mangrove (and a few other marine plants which I named) had lived in the sea, if they had been found only in a fossil state. On another occasion he was almost equally indignant because I rejected with scorn the notion that a continent had formerly extended between Australia and S. America. I have known hardly any man more lovable than Hooker.

A little later I became intimate with Huxley. His mind is as quick as a flash of lightning and as sharp as a razor. He is the best talker whom I have known. He never writes and never says anything flat. From his conversation no one would suppose that he could cut up his opponents in so trenchant a manner as he can do and does do. He has been a most kind friend to me and would always take any trouble for me. He has been the mainstay in England of the principle of the gradual evolution of organic beings. Much splendid work as he has done in Zoology, he would have done far more, if his time had not been so largely consumed by official and literary work, and by his efforts to improve the education of the country. He would allow me to say anything to him: many years ago I thought that it was a pity that he attacked so many scientific men, although I believe that he was right in each particular case, and I said so to him. He denied the charge indignantly, and I answered that I was very glad to hear that I was mistaken. We had been talking about his well-deserved attacks on Owen, so I said after a time, "How well you have exposed Ehrenberg's blunders;" he agreed and added that it was necessary for science that such mistakes should be exposed. Again after a time, I added: "Poor Agassiz has fared ill under your hands." Again I added another name, and now his bright eyes flashed on me, and he burst out laughing, anathematising me in some manner. He is a splendid man and has worked well for the good of mankind.

I may here mention a few other eminent men whom I have occasionally seen, but I have little to say about them worth saying. I felt a high reverence for Sir J. Herschel, and was delighted to dine with him at his charming house at the C. of Good Hope and afterwards at his London house. I saw him, also, on a few other occasions. He never talked much, but every word which he uttered was worth listening to. He was very shy and he often had a distressed expression. Lady Caroline Bell, at whose house I dined at the C. of Good Hope, admired Herschel much, but said that he always came into a room as if he knew that his hands were dirty, and that he knew that his wife knew that they were dirty.

I once met at breakfast at Sir R. Murchison's house, the illustrious Humboldt, who honoured me by expressing a wish to see me. I was a little disappointed with the great man, but my anticipations probably were too high. I can remember nothing distinctly about our interview, except that Humboldt was very cheerful and talked much.

I used to call pretty often on Babbage and regularly attended his famous evening parties. He was always worth listening to, but he was a disappointed and discontented man; and his expression was often or generally morose. I do not believe that he was half as sullen as he pretended to be. One day he told me that he had invented a plan by which all fires could be effectively stopped, but added, – "I shan't publish it – damn them all, let all their houses be burnt." The all were the inhabitants of London. Another day he told me that he had seen a pump on a road-side in Italy, with a pious inscription on it to the effect that the owner had erected the pump for the love of God and his country, that the tired wayfarer might drink. This led Babbage to examine the pump closely and he soon discovered that every time that a wayfarer pumped some water for himself, he pumped a larger quantity into the owner's house. Babbage then added – "There is only one thing which I hate more than piety, and that is patriotism." But I believe that his bark was much worse than his bite.

Herbert Spencer's conversation seemed to me very interesting, but I did not like him particularly, and did not feel that I could easily have become intimate with him. I think that he was extremely egotistical. After reading any of his books, I generally feel enthusiastic admiration for his transcendent talents, and have often wondered whether in the distant future he would rank with such great men as Descartes, Leibnitz, etc., about whom, however, I know very little. Nevertheless I am not conscious of having profited in my own work by Spencer's writings. His deductive manner of treating every subject is wholly opposed to my frame of mind. His conclusions never convince me: and over and over again I have said to myself, after reading one of his discussions,—"Here would be a fine subject for half-a-dozen years' work." His fundamental generalisations (which have been compared in importance by some persons with Newton's laws!) – which I daresay may be very valuable under a philosophical point of view, are of such a nature that they do not seem to me to be of any strictly scientific use. They partake more of the nature of definitions than of laws of nature. They do not aid one in predicting what will happen in any particular case. Anyhow they have not been of any use to me.

Speaking of H. Spencer reminds me of Buckle, whom I once met at Hensleigh Wedgwood's. I was very glad to learn from him his system of collecting facts. He told me that he bought all the books which he read, and made a full index to each, of the facts which he thought might prove serviceable to him, and that he could always remember in what book he had read anything, for his memory was wonderful. I then asked him how at first he could judge what facts would be serviceable and he answered that he did not know, but that a sort of instinct guided him. From this habit of making indices, he was enabled to give the astonishing number of references on all sorts of subjects, which may be found in his *History of Civilisation*. This book I thought most interesting and read it twice; but I doubt whether his generalisations are worth anything. H. Spencer told me that he had never read a line of it! Buckle was a great talker, and I listened to him without saying hardly a word, nor indeed could I have done so, for he left no gaps. When Effie began to sing, I jumped up and said that I must listen to her. This I suppose offended him, for after I had moved away, he turned round to a friend, and said (as was overheard by my brother), "Well Mr Darwin's books are much better than his conversation." What he really meant was that I did not properly appreciate his conversation.

Of other great literary men, I once met Sydney Smith at Dean Milman's house. There was something inexplicably amusing in every word which he uttered. Perhaps this was partly due to the expectation of being amused. He was talking about Lady Cork, who was then extremely old. This was the lady, who, as he said, was once so much affected by one of his charity sermons, that she <u>borrowed</u> a guinea from a friend to put into the Plate. He now said, "It is generally believed that my dear old friend Lady Cork has been overlooked"; and he said this in such a manner that no one could for a moment doubt that he meant that his dear old friend had been overlooked by the devil. How he managed to express this I know not.

I likewise once met Macaulay at Lord Stanhope's (the historian's) house, and as there was only one other man at dinner, I had a grand opportunity of hearing him converse, and he was very agreeable. He did not talk at all too much; nor indeed could such a man talk too much, as long as he allowed others to turn the stream of his conversation, and this he did allow.

Lord Stanhope once gave me a curious little proof of the accuracy and fullness of Macaulay's memory: many historians used often to meet at Lord Stanhope's house, and, in discussing various subjects, they would sometimes differ from Macaulay, and formerly they often referred to some book to see who was right; but latterly, as Lord Stanhope noticed, no historian ever took this trouble, and whatever Macaulay said was final.

On another occasion I met at Ld. Stanhope's house one of his parties of historians and other literary men, and amongst them were Motley and Grote. After luncheon I walked about Chevening Park for nearly an hour with Grote, and was much interested by his conversation and pleased by the simplicity and absence of all pretension in his manners.

I met another set of great men at breakfast at Ld. Stanhope's house in London. After breakfast was quite over, Monckton Milnes (Ld. Houghton now) walked in, and after looking round, exclaimed – (justifying Sidney Smith's nickname of "the cool of the evening") – "Well, I declare, you are all very premature."

Long ago I dined occasionally with the old Earl Stanhope, the father of the historian. I have heard that his father, the democratic earl, well-known at the time of the French Revolution, had his son educated as a blacksmith, as he declared that every man ought to know some trade. The old Earl, whom I knew, was a strange man, but what little I saw of him, I liked much. He was frank, genial, and pleasant. He had strongly-marked features, with a brown complexion, and his clothes, when I saw him, were all brown. He seemed to believe in everything which was to others utterly incredible. He said one day to me, "Why don't you give up your fiddle-faddle of geology and zoology, and turn to the occult sciences?" The historian (then Ld. Mahon) seemed shocked at such a speech to me, and his charming wife much amused.

The last man whom I will mention is Carlyle, seen by me several times at my brother's house and two or three times at my own house. His talk was very racy and interesting, just like his writings, but he sometimes went on too long on the same subject. I remember a funny dinner at my brother's where, amongst a few others, were Babbage and Lyell, both of whom liked to talk. Carlyle, however, silenced every one by haranguing during the whole dinner on the advantages of silence. After dinner, Babbage, in his grimmest manner, thanked Carlyle for his very interesting Lecture on Silence.

Carlyle sneered at almost every one. One day in my house he called Grote's *History* "a fetid quagmire, with nothing spiritual about it." I always thought, until his *Reminiscences* appeared, that his sneers were partly jokes, but this now seems rather doubtful. His expression was that of a depressed, almost despondent, yet benevolent man; and it is notorious how heartily he laughed. I believe that his benevolence was real, though stained by not a little jealousy. No one can doubt about his extraordinary power of drawing vivid pictures of things and men – far more vivid, as it appears to me, than any drawn by Macaulay. Whether his pictures of men were true ones is another question. He has been all-powerful in impressing some grand moral truths on the minds of men. On the other hand, his views about slavery were revolting. In his eyes might was right. His mind seemed to me a very narrow one; even if all branches of science, which he despised, are excluded. It is astonishing to me that Kingsley should have spoken of him as a man well fitted to advance science. He laughed to scorn the idea that a mathematician, such as Whewell, could judge, as I maintained he could, of Goethe's views on light. He thought it a most ridiculous thing that any one should care whether a glacier moved a little quicker or a little slower, or moved at all. As far as I could judge, I never met a man with a mind so ill adapted for scientific research.

Whilst living in London, I attended as regularly as I could the meetings of several scientific societies, and acted as secretary to the Geological Society. But such attendance, and ordinary society, suited my health so badly that we resolved to live in the country, which we both preferred and have never repented of.* Ω

* CD writes here 'written April 1881'.

Residence at Down from Sep. 14 1842
to the present time 1876

After several fruitless searches in Surrey and elsewhere, we found this house and purchased it. I was pleased with the diversified appearance of the vegetation proper to a chalk district, and so unlike what I had been accustomed to in the Midland counties; and still more pleased with the extreme quietness and rusticity of the place. It is not, however, quite so retired a place as a writer in a German periodical makes it, who says that my house can be approached only by a mule-track! Our fixing ourselves here has answered admirably in one way which we did not anticipate, namely, by being very convenient for frequent visits from our children, who never miss an opportunity of doing so when they can.

Few persons can have lived a more retired life than we have done. Besides short visits to the houses of relations, and occasionally to the seaside and elsewhere, we have gone nowhere. During the first part of our residence we went a little into society, and received a few friends here; but my health almost always suffered from the excitement, violent shivering and vomiting attacks being thus brought on. I have therefore been compelled for many years to give up all dinner-parties; and this has been somewhat of a deprivation to me, as such parties always put me into high spirits. From the same cause I have been able to invite here very few scientific acquaintances. Whilst I was young and strong I was capable of very warm attachments, but of late years, though I still have very friendly feelings towards many persons, I have lost the power of becoming deeply attached to anyone, not even so deeply to my good and dear friends Hooker and Huxley, as I should formerly have been. As far as I can judge this grievous loss of feeling has gradually crept over me, from the expectation of much distress afterwards from exhaustion having become firmly associated in my mind with seeing and talking with anyone for an hour, except my wife and children.

My chief enjoyment and sole employment throughout life has been scientific work; and the excitement from such work makes me for the time forget, or drives quite away, my daily discomfort. I have therefore nothing to record during the rest of my life, except the publication of my several books. Perhaps a few details how they arose may be worth giving.

My Several Publications

In the early part of 1844, my observations on the Volcanic Islands visited during the voyage of the *Beagle* were published. In 1845, I took much pains in correcting a new edition of my *Journal of Researches*, which was originally published in 1839 as part of FitzRoy's work. The success of this my first literary child always tickles my vanity more than that of any of my other books. Even to this day it sells steadily in England and the United States, and has been translated for the second time into German, and into French and other languages. This success of a book of travels, especially of a scientific one, so many years after its first publication, is surprising. Ten thousand copies have now been sold in England of the second edition. In 1846 my *Geological Observations on South America* were published. I record in a little diary, which I have always kept, that my three geological books (*Coral Reefs* included) consumed four and a half years' steady work; "and now it is ten years since my return to England. How much time have I lost by illness?" I have nothing to say about these three books except that to my surprise new editions have lately been called for.

In October, 1846, I began to work on Cirripedia. When on the coast of Chile, I found a most curious form, which burrowed into the shells of Concholepas, and which differed so much from all other Cirripedes that I had to form a new sub-order for its sole reception. Lately an allied burrowing genus has been found on the shores of Portugal. To understand the structure of my new Cirripede I had to examine and dissect many of the common forms: and this gradually led me on to take up the whole group. I worked steadily on the subject for the next eight years, and ultimately published two thick volumes, describing all the known living species, and two thin quartos on the extinct species. I do not doubt that Sir E. Lytton Bulwer had me in his mind when he introduces in one of his novels a Professor Long, who had written two huge volumes on Limpets.

Although I was employed during eight years on this work, yet I record in my diary that about two years out of this time was lost by illness. On this account I went in 1848 for some months to Malvern for hydropathic treatment, which did me much good, so that on my return home I was able to resume work. So much was I out of health that when my dear father died on November 13th, 1847,* I was unable to attend his funeral or to act as one of his executors.

My work on the Cirripedia possesses, I think, considerable value, as besides describing several new and remarkable forms, I made out the homologies of the various parts – I discovered the cementing apparatus, though I blundered dreadfully about the cement glands – and lastly I proved the existence in certain genera of minute males complemental to and parasitic on the hermaphrodites. This latter discovery has at last been fully confirmed; though at one time a German writer was pleased to attribute the whole account to my fertile imagination. The Cirripedes form a highly varying and difficult group of species to class; and my work was of considerable use to me, when I had to discuss in the *Origin of Species* the principles of a natural classification. Nevertheless, I doubt whether the work was worth the consumption of so much time.

From September 1854 onwards I devoted all my time to arranging my huge pile of notes, to observing, and experimenting, in relation to the transmutation of species. During the voyage of the *Beagle* I had been deeply impressed by discovering in the Pampean formation great fossil animals covered with armour like that on the existing armadillos; secondly, by the manner in which closely allied animals replace one another in proceeding southwards over the Continent; and thirdly, by the South American character of most of the productions of the Galapagos archipelago, and more especially by the manner in which they differ

* CD famously misdates his father's death as 1847, not 1848.

slightly on each island of the group; none of these islands appearing to be very ancient in a geological sense.

It was evident that such facts as these, as well as many others, could be explained on the supposition that species gradually become modified; and the subject haunted me. But it was equally evident that neither the action of the surrounding conditions, nor the will of the organisms (especially in the case of plants), could account for the innumerable cases in which organisms of every kind are beautifully adapted to their habits of life, – for instance, a woodpecker or tree-frog to climb trees, or a seed for dispersal by hooks or plumes. I had always been much struck by such adaptations, and until these could be explained it seemed to me almost useless to endeavour to prove by indirect evidence that species have been modified.

After my return to England it appeared to me that by following the example of Lyell in Geology, and by collecting all facts which bore in any way on the variation of animals and plants under domestication and nature, some light might perhaps be thrown on the whole subject. My first note-book was opened in July 1837. I worked on true Baconian principles, and without any theory collected facts on a wholesale scale, more especially with respect to domesticated productions, by printed enquiries, by conversation with skilful breeders and gardeners, and by extensive reading. When I see the list of books of all kinds which I read and abstracted, including whole series of Journals and Transactions, I am surprised at my industry. I soon perceived that selection was the keystone of man's success in making useful races of animals and plants. But how selection could be applied to organisms living in a state of nature remained for some time a mystery to me.

In October 1838, that is, fifteen months after I had begun my system-atic enquiry, I happened to read for amusement 'Malthus on *Population*', and being well prepared to appreciate the struggle for existence which everywhere goes on from long-continued observation of the habits of animals and plants, it at once struck me that under these circumstances favourable variations would tend to be preserved, and unfavourable ones to be destroyed. The result of this would be the formation of new species. Here, then, I had at last got a theory by which to work; but I was so anxious to avoid prejudice, that I determined not for some time to write even the briefest sketch of it. In June 1842 I first allowed myself the satisfaction of writing a very brief abstract of my theory in pencil in 35 pages; and this was enlarged during the summer of 1844 into one of 230 pages, which I had fairly copied out and still possess.

But at that time I overlooked one problem of great importance; and it is astonishing to me, except on the principle of Columbus and his egg, how I could have overlooked it and its solution. This problem is the tendency in organic beings descended from the same stock to diverge in character as they become modified. That they have diverged greatly is obvious from the manner in which species of all kinds can be classed under genera, genera under families, families under sub-orders, and so forth; and I can remember the very spot in the road, whilst in my carriage, when to my joy the solution occurred to me; and this was long after I had come to Down. The solution, as I believe, is that the modified offspring of all dominant and increasing forms tend to become adapted to many and highly diversified places in the economy of nature.

Early in 1856 Lyell advised me to write out my views pretty fully, and I began at once to do so on a scale three or four times as extensive as that which was afterwards followed in my Origin of Species; yet it was only an abstract of the materials which I had collected, and I got through about half the work on this scale. But my plans were overthrown, for early in the summer of 1858 Mr Wallace, who was then in the Malay archipelago, sent me an essay on the Tendency of Varieties to depart indefinitely from the Original Type; and this essay contained exactly the same theory as mine. Mr Wallace expressed the wish that if I thought well of his essay, I should send it to Lyell for perusal.

The circumstances under which I consented at the request of Lyell and Hooker to allow of an extract from my MS., together with a letter to Asa Gray, dated September 5, 1857, to be published at the same time with Wallace's Essay, are given in the *Journal of the Proceedings of the Linn: Soc.*, 1858, p. 45. I was at first very unwilling to consent, as I thought Mr Wallace might consider my doing so unjustifiable, for I did not then know how generous and noble was his disposition. The extract from my MS. and the letter to Asa Gray had neither been intended for publication, and were badly written. Mr Wallace's essay, on the other hand, was admirably expressed and quite clear. Nevertheless, our joint productions excited very little attention, and the only published notice of them which I can remember was by Professor Haughton of Dublin, whose verdict was that all that was new in them was false, and what was true was old. This shows how necessary it is that any new view should be explained at considerable length in order to arouse public attention.

In September 1858 I set to work by the strong advice of Lyell and Hooker to prepare a volume on the transmutation of species, but was often interrupted by ill-health, and short visits to Dr. Lane's delightful hydropathic establishment at Moor Park. I abstracted the MS. begun on a much larger scale in 1856, and completed the volume on the same reduced scale. It cost me thirteen months and ten days' hard labour. It was published under the title of the *Origin of Species*, in November 1859. Though considerably added to and corrected in the later editions, it has remained substantially the same book.

It is no doubt the chief work of my life. It was from the first highly successful. The first small edition of 1250 copies was sold on the day of publication, and a second edition of 3000 copies soon afterwards. Sixteen thousand copies have now (1876) been sold in England and considering how stiff a book it is, this is a large sale. It has been translated into almost every European tongue, even into such languages as Spanish, Bohemian, Polish, and Russian. ¶ It has also, according to Miss Bird, been translated into Japanese, and is there much studied. Ω Even an essay in Hebrew has appeared on it, showing that the theory is contained in the Old Testament! The reviews were very numerous; for a time I collected all that appeared on the *Origin* and on my related books, and these amount (excluding newspaper reviews) to 265; but after a time I gave up the attempt in despair. Many separate essays and books on the subject have appeared; and in Germany a catalogue or bibliography on "Darwinismus" has appeared every year or two.

The success of the *Origin* may, I think, be attributed in large part to my having long before written two condensed sketches, and to my having finally abstracted a much larger manuscript, which was itself an abstract. By this means I was enabled to select the more striking facts and conclusions. I had, also, during many years, followed a golden rule, namely, that whenever a published fact, a new observation or thought came across me, which was opposed to my general results, to make a memorandum of it without fail and at once; for I had found by experience that such facts and thoughts were far more apt to escape from the memory than favourable ones. Owing to this habit, very few objections were raised against my views which I had not at least noticed and attempted to answer.

It has sometimes been said that the success of the *Origin* proved "that the subject was in the air," or "that men's minds were prepared for it." I do not think that this is strictly true, for I occasionally sounded not a few naturalists, and never happened to come across a single one who seemed to doubt about the permanence of species. Even Lyell and Hooker, though they would listen with interest to me, never seemed to agree. I tried once or twice to explain to able men what I meant by natural selection, but signally failed. What I believe was strictly true is that innumerable well-observed facts were stored in the minds of naturalists, ready to take their proper places as soon as any theory which would receive them was sufficiently explained. Another element in the success of the book was its moderate size; and this I owe to the appearance of Mr Wallace's essay; had I published on the scale in which I began to write in 1856, the book would have been four or five times as large as the Origin, and very few would have had the patience to read it.

I gained much by my delay in publishing from about 1839, when the theory was clearly conceived, to 1859; and I lost nothing by it, for I cared very little whether men attributed most originality to me or Wallace; and his essay no doubt aided in the reception of the theory. I was forestalled in only one important point, which my vanity has always made me regret, namely, the explanation by means of the Glacial period of the presence of the same species of plants and of some few animals on distant mountain summits and in the arctic regions. This view pleased me so much that I wrote it out in extenso, and* it was read by Hooker

* CD deleted in pencil 'I believe this'.

some years before E. Forbes published his celebrated memoir on the subject. In the very few points in which we differed, I still think that I was in the right. I have never, of course, alluded in print to my having independently worked out this view.

Hardly any point gave me so much satisfaction when I was at work on the *Origin*, as the explanation of the wide difference in many classes between the embryo and the adult animal, and of the close resemblance of the embryos within the same class. No notice of this point was taken, as far as I remember, in the early reviews of the *Origin*, and I recollect expressing my surprise on this head in a letter to Asa Gray. Within late years several reviewers have given the whole credit of the idea to Fritz Müller and Haeckel, who undoubtedly have worked it out much more fully, and in some respects more correctly than I did. I had materials for a whole chapter on the subject, and I ought to have made the discussion longer; for it is clear that I failed to impress my readers; and he who succeeds in doing so deserves, in my opinion, all the credit.

This leads me to remark that I have almost always been treated honestly by my reviewers, passing over those without scientific knowledge as not worthy of notice. My views have often been grossly misrepresented, bitterly opposed and ridiculed, but this has been generally done, as I believe, in good faith. I must, however, except Mr Mivart, who as an American expressed it in a letter has acted towards me "like a pettifogger", or as Huxley has said "like an Old Bailey lawyer." On the whole I do not doubt that my works have been over and over again greatly overpraised. I rejoice that I have avoided controversies, and this I owe to Lyell, who many years ago, in reference to my geological works, strongly advised me never to get entangled in a controversy, as it rarely did any good and caused a miserable loss of time and temper.

¶ Whenever I have found out that I have blundered, or that my work has been imperfect, and when I have been contemptuously criticised, and even when I have been overpraised, so that I have felt mortified, it has been my greatest comfort to say hundreds of times to myself that "I have worked as hard and as well as I could, and no man can do more than this." I remember when in Good Success Bay, in Tierra del Fuego, thinking, (and I believe that I wrote home to the effect) that I could not employ my life better than in adding a little to natural science. This I have done to the best of my abilities, and critics may say what they like, but they cannot destroy this conviction. Ω

During the two last months of the year 1859 I was fully occupied in preparing a second edition of the *Origin*, and by an enormous correspondence. On January 7th, 1860, I began arranging my notes for my work on the *Variation of Animals and Plants under Domestication*; but it was not published until the beginning of 1868; the delay having been caused partly by frequent illnesses, one of which lasted seven months, and partly by having been tempted to publish on other subjects which at the time interested me more.

On May 15 1862, my little book on the *Fertilisation of Orchids*, which cost me ten months' work, was published: most of the facts had been slowly accumulated during several previous years. During the summer of 1839, and, I believe, during the previous summer, I was led to attend to the cross-fertilisation of flowers by the aid of insects, from having come to the conclusion in my speculations on the origin of species, that crossing played an important part in keeping specific forms constant. I attended to the subject more or less during every subsequent summer; and my interest in it was greatly enhanced by having procured and read in November 1841, through the advice of Robert Brown, a copy of C. K. Sprengel's wonderful book, *Das entdeckte Geheimnis der Natur*. For some years before 1862 I had specially attended to the fertilisation of our British orchids; and it seemed to me the best plan to prepare as complete a treatise on this group of plants as well as I could, rather than to utilise the great mass of matter which I had slowly collected with respect to other plants.

My resolve proved a wise one; for since the appearance of my book, a surprising number of papers and separate works on the fertilisation of all kinds of flowers have appeared; and these are far better done than I could possibly have effected. The merits of poor old Sprengel, so long overlooked, are now fully recognised many years after his death.

During this same year I published in the *Journal of the Linnean Society*, a paper 'On the Two Forms, or Dimorphic Condition of

Primula', and during the next five years, five other papers on dimorphic and trimorphic plants. I do not think anything in my scientific life has given me so much satisfaction as making out the meaning of the structure of these plants. I had noticed in 1838 or 1839 the dimorphism of *Linum flavum*, and had at first thought that it was merely a case of unmeaning variability. But on examining the common species of Primula, I found that the two forms were much too regular and constant to be thus viewed. I therefore became almost convinced that the common cowslip and primrose were on the high-road to became diœcious; – that the short pistil in the one form, and the short stamens in the other form were tending towards abortion. The plants were therefore subjected under this point of view to trial; but as soon as the flowers with short pistils fertilised with pollen from the short stamens, were found to yield more seeds than any other of the four possible unions, the abortion-theory was knocked on the head. After some additional experiment, it became evident that the two forms, though both were perfect hermaphrodites, bore almost the same relation to one another as do the two sexes of an ordinary animal. With Lythrum we have the still more wonderful case of three forms standing in a similar relation to one another. I afterwards found that the offspring from the union of two plants belonging to the same forms presented a close and curious analogy with hybrids from the union of two distinct species.

In the autumn of 1864 I finished a long paper on Climbing Plants, and sent it to the Linnean Society. The writing of this paper cost me four months: but I was so unwell when I received the proof-sheets that I was forced to leave them very badly and often obscurely expressed. The paper was little noticed, but when in 1875 it was corrected and published as a separate book it sold well. I was led to take up this subject by reading a short paper by Asa Gray, published in 1858, on the movements of the tendrils of a Cucurbitacean plant. He sent me seeds, and on raising some plants I was so much fascinated and perplexed by the revolving movements of the tendrils and stems, which movements are really very simple, though appearing at first very complex, that I procured various other kinds of Climbing Plants, and studied the whole subject. I was all the more attracted to it, from not being at all satisfied with the explanation which Henslow gave us in his Lectures, about Twining plants, namely, that they had a natural tendency to grow up a spire. This explanation proved quite erroneous. Some of the adaptations displayed by climbing plants are as beautiful as those by Orchids for ensuring cross-fertilisation.

My *Variation of Animals and Plants under Domestication* was begun, as already stated, in the beginning of 1860, but was not published until the beginning of 1868. It is a big book, and cost me four years and two months' hard labour. It gives all my observations and an immense number of facts collected from various sources, about our domestic productions. In the second volume the causes and laws of variation, inheritance, &c., are discussed, as far as our present state of knowledge permits. Towards the end of the work I give my well-abused hypothesis of Pangenesis. An unverified hypothesis is of little or no value. But if any one should hereafter be led to make observations by which some such hypothesis could be established, I shall have done good service, as an astonishing number of isolated facts can thus be connected together and rendered intelligible. In 1875 a second and largely corrected edition, which cost me a good deal of labour, was brought out.

My *Descent of Man* was published in Feb. 1871. As soon as I had become, in the year 1837 or 1838, convinced that species were mutable productions, I could not avoid the belief that man must come under the same law. Accordingly I collected notes on the subject for my own satisfaction, and not for a long time with any intention of publishing. Although in the *Origin of Species*, the derivation of any particular species is never discussed, yet I thought it best, in order that no honourable man should accuse me of concealing my views, to add that by the work in question "light would be thrown on the origin of man and his history." It would have been useless and injurious to the success of the book to have paraded without giving any evidence my conviction with respect to his origin.

But when I found that many naturalists fully accepted the doctrine of the evolution of species, it seemed to me advisable to work up such notes as I possessed and to publish a special treatise on the origin of man. I was the more glad to do so, as it gave me an opportunity of fully discussing sexual selection, – a subject which had always greatly interested me. This subject, and that of the variation of our domestic productions, together with the causes and laws of variation, inheritance, etc., and the intercrossing of Plants, are the sole subjects which I have been able to write about in full, so as to use all the materials which I had collected. The *Descent of Man* took me three years to write, but then as usual some of this time was lost by ill health, and some was consumed by preparing new editions and other minor works. A second and largely corrected edition of the Descent appeared in 1874.

My book on the *Expression of the Emotions in Men and Animals* was published in the autumn of 1872. I had intended to give only a chapter on the subject in the *Descent of Man*, but as soon as I began to put my notes together, I saw that it would require a separate Treatise. My first child was born on December 27th, 1839, and I at once commenced to make notes on the first dawn of the various expressions which he exhibited, for I felt convinced, even at this early period, that the most complex and fine shades of expression must all have had a gradual and natural origin. During the summer of the following year, 1840, I read Sir C. Bell's admirable work on Expression, and this greatly increased the interest which I felt in the subject, though I could not at all agree with his belief that various muscles had been specially created for the sake of expression. From this time forward I occasionally attended to the subject, both with respect to man and our domesticated animals. My book sold largely; 5267 copies having been disposed of on the day of publication.

In the summer of 1860 I was idling and resting near Hartfield, where two species of Drosera abound; and I noticed that numerous insects had been entrapped by the leaves. I carried home some plants, and on giving them insects saw the movements of the tentacles, and this made me think it probable that the insects were caught for some special purpose. Fortunately a crucial test occurred to me, that of placing a large number of leaves in various nitrogenous and non-nitrogenous fluids of equal density; and as soon as I found that the former alone excited energetic movements, it was obvious that here was a fine new field for investigation.

During subsequent years, whenever I had leisure, I pursued my experiments, and my book on *Insectivorous Plants* was published July 1875, – that is sixteen years after my first observations. The delay in this case, as with all my other books, has been a great advantage to me; for a man after a long interval can criticise his own work, almost as well as if it were that of another person. The fact that a plant should secrete, when properly excited, a fluid containing an acid and ferment, closely analogous to the digestive fluid of an animal, was certainly a remarkable discovery.

During this autumn of 1876 I shall publish on the *Effects of Cross- and Self-Fertilisation in the Vegetable Kingdom*. This book will form a complement to that on the *Fertilisation of Orchids*, in which I showed how perfect were the means for cross-fertilisation, and here I shall show how important are the results. I was led to make, during eleven years, the numerous experiments recorded in this volume, by a mere accidental observation; and indeed it required the accident to be repeated before my attention was thoroughly aroused to the remarkable fact that seedlings of self-fertilised parentage are inferior, even in the first generation, in height and vigour to seedlings of cross-fertilised parentage. I hope also to republish a revised edition of my book on Orchids, and hereafter my papers on dimorphic and trimorphic plants, together with some additional observations on allied points which I never have had time to arrange. My strength will then probably be exhausted, and I shall be ready to exclaim "Nunc dimittis." ¶ *The Effects of Cross- and Self-Fertilisation* was published in the autumn of 1876; and the results there arrived at explain, as I believe, the endless and wonderful contrivances for the transportal of pollen from one plant to another of the same species. I now believe, however, chiefly from the observations of Hermann Müller, that I ought to have insisted more strongly than I did on the many adaptations for self-fertilisation; though I was well aware of many such adaptations. A much enlarged edition of my *Fertilisation of Orchids* was published in 1877.

In this same year *The Different Forms of Flowers*, etc., appeared, and in 1880 a second edition. This book consists chiefly of the several papers on heterostyled flowers, originally published by the Linnean Society, corrected, with much new matter added, together with observations on some other cases in which the same plant bears two kinds of flowers. As before remarked, no little discovery of mine ever gave me so much pleasure as the making out the meaning of heterostyled flowers. The results of crossing such flowers in an illegitimate manner, I believe to be very important as bearing on the sterility of hybrids; although these results have been noticed by only a few persons.

In 1879, I had a translation of Dr. Ernst Krause's *Life of Erasmus Darwin* published, and I added a sketch of his character and habits from materials in my possession. Many persons have been much interested by this little life, and I am surprised that only 800 or 900 copies were sold. Owing to my having accidentally omitted to mention that Dr. Krause had enlarged and corrected his article in German before it was translated, Mr Samuel Butler abused me with almost insane virulence. How I offended him so bitterly, I have never been able to understand. The subject gave rise to some controversy in the Athenæum newspaper and Nature. I laid all the documents before some good judges, viz. Huxley, Leslie Stephen, Litchfield, etc., and they were all unanimous that the attack was so baseless that it did not deserve any public answer; for I had already expressed privately my regret to Mr. Butler for my accidental omission. Huxley consoled me by quoting some German lines from Goethe, who had been attacked by someone, to the effect "that every Whale has its Louse."

In 1880 I published, with Frank's assistance, our *Power of Movement in Plants*. This was a tough piece of work. The book bears somewhat the same relation to my little book on *Climbing Plants*, which *Cross-Fertilisation* did to the *Fertilisation of Orchids*; for in accordance with the principles of evolution it was impossible to account for climbing plants having been developed in so many widely different groups, unless all kinds of plants possess some slight power of movement of an analogous kind. This I proved to be the case, and I was further led to a rather wide generalisation, viz., that the great and important classes of movements, excited by light, the attraction of gravity, &c., are all modified forms of the fundamental movement of circumnutation. It has always pleased me to exalt plants in the scale of organised beings; and I therefore felt an especial pleasure in showing how many and what admirably well adapted movements the tip of a root possesses.

I have now (May 1, 1881) sent to the printers the MS. of a little book on The Formation of Vegetable Mould through the Action of Worms. This is a subject of but small importance; and I know not whether it will interest any readers, but it has interested me. It is the completion of a short paper read before the Geological Society more than 40 years ago, and has revived old geological thoughts.*Ω

I have now mentioned all the books which I have published, and these have been the mile-stones in my life, so that little remains to be said. I am not conscious of any change in my mind during the last thirty years, excepting in one point presently to be mentioned; nor indeed could any change have been expected unless one of general deterioration. But my father lived to his eighty-third year with his mind as lively as ever it was, and all his faculties undimmed; and I hope that I may die before my mind fails to a sensible extent. I think that I have become a little more skilful in guessing right explanations and in devising experimental tests; but this may probably be the result of mere practice, and of a larger store of knowledge. I have as much difficulty as ever in expressing myself clearly and concisely; and this difficulty has caused me a very great loss of time; but it has had the compensating advantage of forcing me to think long and intently about every sentence, and thus I have been often led to see errors in reasoning and in my own observations or those of others.

There seems to be a sort of fatality in my mind leading me to put at first my statement and proposition in a wrong or awkward form. Formerly I used to think about my sentences before writing them down; but for several years I have found that it saves time to scribble in a vile hand whole pages as quickly as I possibly can, contracting half the words; and then correct deliberately. Sentences thus scribbled down are often better ones than I could have written deliberately.

Having said this much about my manner of writing, I will add that with my larger books I spend a good deal of time over the general

* CD dates this 'May 1 1881'.

arrangement of the matter. I first make the rudest outline in two or three pages, and then a larger one in several pages, a few words or one word standing for a whole discussion or series of facts. Each of these headings is again enlarged and often transformed before I begin to write in extenso. As in several of my books facts observed by others have been very extensively used, and as I have always had several quite distinct subjects in hand at the same time, I may mention that I keep from 30 to 40 large portfolios, in cabinets with labelled shelves, into which I can at once put a detached reference or memorandum. I have bought many books and at their ends I make an index of all the facts that concern my work; or, if the book is not my own, write out a separate abstract, and of such abstracts I have a large drawer full. Before beginning on any subject I look to all the short indexes and make a general and classified index, and by taking the one or more proper portfolios I have all the information collected during my life ready for use.

I have said that in one respect my mind has changed during the last 20 or 30 years. Up to the age of thirty, or beyond it, poetry of many kinds, such as the works of Milton, Gray, Byron, Wordsworth, Coleridge, and Shelley, gave me great pleasure, and even as a schoolboy I took intense delight in Shakespeare, especially in the historical plays. I have also said that formerly pictures gave me considerable, and music very great delight. But now for many years I cannot endure to read a line of poetry: I have tried lately to read Shakespeare, and found it so intolerably dull that it nauseated me. I have also almost lost any taste for pictures or music. – Music generally sets me thinking too energetically on what I have been at work on, instead of giving me pleasure. I retain some taste for fine scenery, but it does not cause me the exquisite delight which it formerly did. On the other hand, novels which are works of the imagination, though not of a very high order, have been for years a wonderful relief and pleasure to me, and I often bless all novelists. A surprising number have been read aloud to me, and I like all if moderately good, and if they do not end unhappily – against which a law ought to be passed. A novel, according to my taste, does not come into the first class unless it contains some person whom one can thoroughly love, and if it be a pretty woman all the better.

This curious and lamentable loss of the higher aesthetic tastes is all the odder, as books on history, biographies and travels (independently of any scientific facts which they may contain), and essays on all sorts of subjects interest me as much as ever they did. My mind seems to have become a kind of machine for grinding general laws out of large collections of facts, but why this should have caused the atrophy of that part of the brain alone, on which the higher tastes depend, I cannot conceive. A man with a mind more highly organised or better constituted than mine, would not I suppose have thus suffered; and if I had to live my life again I would have made a rule to read some poetry and listen to some music at least once every week; for perhaps the parts of my brain now atrophied could thus have been kept active through use. The loss of these tastes is a loss of happiness, and may possibly be injurious to the intellect, and more probably to the moral character, by enfeebling the emotional part of our nature.

My books have sold largely in England, have been translated into many languages, and passed through several editions in foreign countries. I have heard it said that the success of a work abroad is the best test of its enduring value. I doubt whether this is at all trustworthy; but judged by this standard my name ought to last for a few years. Therefore it may be worth while for me to try to analyse the mental qualities and the conditions on which my success has depended; though I am aware that no man can do this correctly.

I have no great quickness of apprehension or wit which is so remarkable in some clever men, for instance Huxley. ¶ I am therefore a poor critic: a paper or book, when first read, generally excites my admiration, and it is only after considerable reflection that I perceive the weak points. Ω My power to follow a long and purely abstract train of thought is very limited; I should, moreover, never have succeeded with metaphysics or mathematics. My memory is extensive, yet hazy: it suffices to make me cautious by vaguely telling me that I have observed or read something opposed to the conclusion which I am drawing, or on the other hand in favour of it; and after a time I can generally recollect where to search for my authority. So poor in one sense is my memory, that I have never been able to remember for more than a few days a single date or a line of poetry.

Some of my critics have said, "Oh, he is a good observer, but has no power of reasoning." I do not think that this can be true, for the *Origin of Species* is one long argument from the beginning to the end, and it has convinced not a few able men. No one could have written it without having some power of reasoning. I have a fair share of invention and of common sense or judgment, such as every fairly successful lawyer or doctor must have, but not I believe, in any higher degree.

On the favourable side of the balance, I think that I am superior to the common run of men in noticing things which easily escape attention, and in observing them carefully. My industry has been nearly as great as it could have been in the observation and collection of facts. What is far more important, my love of natural science has been steady and ardent. This pure love has, however, been much aided by the ambition to be esteemed by my fellow naturalists. From my early youth I have had the strongest desire to understand or explain whatever I observed, – that is, to group all facts under some general laws. These causes combined have given me the patience to reflect or ponder for any number of years over any unexplained problem. As far as I can judge, I am not apt to follow blindly the lead of other men. I have steadily endeavoured to keep my mind free, so as to give up any hypothesis, however much beloved (and I cannot resist forming one on every subject), as soon as facts are shown to be opposed to it. Indeed I have had no choice but to act in this manner, for with the exception of the Coral Reefs, I cannot remember a single first-formed hypothesis which had not after a time to be given up or greatly modified. This has naturally led me to distrust greatly deductive reasoning in the mixed sciences. On the other hand, I am not very sceptical, – a frame of mind which I believe to be injurious to the progress of science; a good deal of scepticism in a scientific man is ¶ advisable to avoid much loss of time; for I have met with not a few men, who I feel sure have often thus been deterred from experiment or observations, which would have proved directly or indirectly serviceable.

In illustration, I will give the oddest case which I have known. A gentleman (who, as I afterwards heard, was a good local botanist) wrote to me from the Eastern countries that the seeds or beans of the common fieldbean had this year everywhere grown on the wrong side of the pod. I wrote back, asking for further information, as I did not understand what was meant; but I did not receive any answer for a long time. I then saw in two newspapers, one published in Kent and the other in Yorkshire, paragraphs stating that it was a most remarkable fact that "the beans this year had all grown on the wrong side." So I thought that there must be some foundation for so general a statement. Accordingly, I went to my gardener, an old Kentish man, and asked him whether he had heard anything about it; and he answered, "Oh, no, Sir, it must be a mistake, for the beans grow on the wrong side only on Leap-year, and this is not Leap-year." I then asked him how they grew on common years and how on Leap-years, but soon found out that he knew absolutely nothing of how they grew at any time; but he stuck to his belief.

After a time I heard from my first informant, who, with many apologies, said that he should not have written to me had he not heard the statement from several intelligent farmers; but that he had since spoken again to every one of them, and not one knew in the least what he had himself meant. So that here a belief – if indeed a statement with no definite idea attached to it can be called a belief – had spread over almost the whole of England without any vestige of evidence.

I have known in the course of my life only three intentionally falsified statements, and one of these may have been a hoax (and there have been several scientific hoaxes) which, however, took in an American agricultural journal. It related to the formation in Holland of a new breed of oxen by the crossing of distinct species of Bos (some of which I happen to know are sterile together), and the author had the impudence to state that he had corresponded with me, and that I had been deeply impressed with the importance of his results. The article was sent to me by the editor of an English Agricult. Journal, asking for my opinion before republishing it.

A second case was an account of several varieties raised by the author from several species of Primula, which had spontaneously yielded a full complement of seed, although the parent plants had been carefully protected from the access of insects. This account was published before I had discovered the meaning of heterostylism, and the whole statement must have been fraudulent, or there was neglect in excluding insects so gross as to be scarcely credible.

The third case was more curious: Mr Huth published in his book on Consanguineous Marriage some long extracts from a Belgian author, who stated that he had interbred rabbits in the closest manner for very many generations without the least injurious effects. The account was published in a most respectable Journal, that of the Royal Medical Soc. of Belgium; but I could not avoid feeling doubts, – I hardly know why, except that there were no accidents of any kind, and my experience in breeding animals made me think this improbable.

So with much hesitation I wrote to Prof. Van Beneden asking him whether the author was a trustworthy man. I soon heard in answer that the Society had been greatly shocked by discovering that the whole account was a fraud. The writer had been publicly challenged in the Journal to say where he had resided and kept his large stock of rabbits while carrying on his experiments, which must have consumed several years, and no answer could be extracted from him. I informed poor Mr Huth, that the account which formed the cornerstone of his argument was fraudulent; and he in the most honourable manner immediately had a slip printed to this effect to be inserted in all future copies of his book which might be sold. Ω

My habits are methodical, and this has been of not a little use for my particular line of work. Lastly, I have had ample leisure from not having to earn my own bread. Even ill-health, though it has annihilated several years of my life, has saved me from the distractions of society and amusement.

Therefore, my success as a man of science, whatever this may have amounted to, has been determined, as far as I can judge, by complex and diversified mental qualities and conditions. Of these the most important have been – the love of science – unbounded patience in long reflecting over any subject – industry in observing and collecting facts – and a fair share of invention as well as of common-sense. With such moderate abilities as I possess, it is truly surprising that thus I should have influenced to a considerable extent the beliefs of scientific men on some important points.

Aug 3rd 1876

This sketch of my life was begun about May 28th. at Hopedene, and since then I have written for nearly an hour on most afternoons.

BIOGRAPHICAL REGISTER

W. F. A. Ainsworth (1807–96) Geologist and surgeon. As a geologist of the Wernerian school, Ainsworth would have believed in the geological primacy of water as an agent of geological change: older rocks such as granite would have been solidified out of an original and universal ocean. As waters receded, sedimentary rocks would have been precipitated at a higher level in the geological column.

Charles Babbage (1792–1871) English mathematician. Pioneer in the design of mechanical computers and one of the founders of the British Association for the Advancement of Science (1831).

Sir Charles Bell (1774–1842) Scottish anatomist and surgeon. Bell carried out pioneering research on the human nervous system. Darwin was particularly interested in Bell's writings on expression, although he disagreed with Bell's view that various human muscles used in expression were unique and had been designed by God.

Robert Brown (1773–1858) Botanist. Brown was keeper of the botanical collections at the British Museum from the late 1820s to the year of his death. Admired by Alexander Von Humboldt (Brown, however, found Humboldt's writing over-wordy and declamatory) and much praised by Darwin.

Dr Samuel Butler (1774–1839) Headmaster of Shrewsbury School for thirty-eight years. Butler was the incarnation of a reliable, mildly innovative public school headmaster. That Darwin should choose to comment upon him in unusually critical terms in the autobiography becomes all the more striking.

Thomas Carlyle (1795–1881) Essayist and historian. Works include *Sartor Resartus/The Tailor Retailored* (1833–4), *The French Revolution* (1837) and *Chartism* (1839). Darwin remembers Carlyle for his sneering and makes one of his few jokes in the autobiography when describing Carlyle's pomposity.

Emma Darwin (née Wedgwood) (1808–96) Darwin's first cousin and his wife. They married in 1839 and had ten children, of whom seven survived to adulthood. Emma was the daughter of Josiah Wedgwood the second. She is characterized as impeccable, the perfect wife and mother, although Darwin was aware that some of his views on Christianity and also some of his judgements on the character of his contemporaries upset her.

Erasmus Darwin (1731–1802) Darwin's grandfather. Erasmus was a famous Enlightenment combination of physician, botanist and poet. His works include *Zoonomia, or the Laws of Organic Life* (1794–6), *The Botanic Garden* (1789–91) and *The Temple of Nature or the Origin of Society* (1803). His speculative views on evolution, sexual life and technology were generally disapproved of in the Victorian age and Charles came to judge much of his scientific writing as speculative and ineffectual.

Erasmus Alvey Darwin (1804–81) Darwin's elder and only brother. Like Charles, Erasmus went to Shrewsbury School and Christ's College, Cambridge. He qualified in medicine at Edinburgh University in 1826 but never practised. A *litterateur*, lifelong bachelor and melancholic, he moved in London literary society and was a friend of Thomas and Jane Carlyle: the latter always found Erasmus (usually called 'Ras') dependable when helping to escort her to the right shops on time. Much of Charles Darwin's intense protectiveness towards family members surfaced again with his elder brother. Even the mildest remarks – hints – by Thomas Carlyle on Ras's character were disapproved of by Charles and receive a curt dismissal in the autobiography.

Dr Robert Waring Darwin (1766–1848) Darwin's father. A physician with an extensive practice in the Shrewsbury area.

Robert FitzRoy (1805–65) Naval officer, meteorologist, geologist and hydrographer. FitzRoy commanded the *Beagle* between 1828 and 1836 and published his account of that voyage in 1839. Governor of New Zealand 1843–5.

He was chief of the meteorological department of the Board of Trade from 1854 and remained a committed Christian throughout his life. He felt various disappointments in his career, especially for being underrecognized as an innovative meteorologist.

Edward Forbes (1815-54) English zoologist, botanist and palaeontologist. Published *A History of British Star-fishes* (1841), *A History of British Mollusca and their Shells*, 4 volumes (1848-53), along with other important geological, botanical and palaeontological papers. He was President of the Geological Society and became Professor of Natural History at Edinburgh University in 1854. Forbes was a gifted scientist who died prematurely, six months after his Edinburgh appointment.

Sir Francis Galton (1822-1911) English explorer, anthropologist and statistician. Galton was Darwin's cousin and strongly believed in the application of the 'truths' of natural selection and evolution to human social questions. He therefore made extensive statistical studies as well as exploring the nature of heredity in the natural order. He coined the word 'eugenics' in 1883 to describe his various scientific activities. Works included *Hereditary Genius* (1869) and *Natural Inheritance* (1889). Darwin was particularly impressed by the evidence for the hereditarian transmission of innate qualities collected in the former work and by implication his own place in that story.

Robert Grant (1793-1874) Physician and zoologist. Grant has an intriguing place in the history of Darwin's development. He was an early proponent of the transmutation of species and would have spoken to Darwin about these matters while he was a medical student in Edinburgh. From 1827 to his death in 1874 he was Professor of Comparative Anatomy and Zoology at University College London, where his unusual and radical views would have continued to have an influence on many subsequent generations of students. In recent Darwin scholarship a persuasive case has been made for Grant as the skeleton in Darwin's scientific cupboard, a presence only made stronger by Darwin's over-defensive dismissal of Grant's time at UCL.

Asa Gray (1810-88) America's leading botanist of the mid-nineteenth century. A Professor of Natural History at Harvard, he was Darwin's strongest early supporter in the United States. In 1857 he was the third scientist after Hooker and Lyell in whom Darwin confided regarding his theory of evolution. While firmly supporting Darwinian evolution, Gray attempted to keep a place for divine action in the actual process of natural selection.

Ernst Haeckel (1834–1919) German professor, zoologist and morphologist. Haeckel was an influential figure in Germany and famous for his theory that 'ontogeny recapitulates phylogeny'. Darwin found the scale of Haeckel's enthusiasm for his theory both flattering and extravagant and was wary of its materialist implications for broader social views.

John Stevens Henslow (1796–1861) Clergyman, botanist, mineralogist and renovator of the botanic garden in Cambridge. Henslow was Darwin's friend as well as teacher and in the autobiography is a model of the courteous and modest clergyman-scientist.

Sir John Herschel (1792–1871) Astronomer, mathematician and philosopher. Herschel was the embodiment of the grandeur that was astronomy, the most epic and religiously instructive of the sciences in the early Victorian age.

Sir Joseph Hooker (1817–1911) English botanist and explorer. Hooker travelled to the Antarctic and then in the late 1840s to the Himalayas. He worked at the Royal Botanic Gardens in Kew from 1855 until 1885, the last twenty years as director. Extremely knowledgeable on taxonomy as well as the geography of plants, he was a genuinely close friend and – of prime importance to Darwin – an ally.

Leonard Horner (1785–1864) Geologist, educationist and one of the founders of the influential *Edinburgh Review*. A sometime warden of University College London and president of the Geological Society of London. One of Horner's daughters married the geologist Sir Charles Lyell.

F. W. H. Alexander von Humboldt (1769–1859) Humboldt was a highly respected, indeed almost legendary German naturalist and traveller. His *Personal Narrative of a Journey to the Equinoctial Regions of the New Continent* (1814–29; 2nd trans. 1851) set a new framework for a total natural history of the earth's zoogeographical and climatic regions. Not least, Humboldt set an example as to how the traveller had to combine exploratory courage with highly detailed collection of information based on personal observation.

T. H. Huxley (1825-95) Naturalist, physiologist, comparative anatomist and educational reformer. Often remembered simply as Darwin's 'bulldog', Huxley had a long and active London-based career, seeking the extension of science education and the teaching of new forms of knowledge. His complex personal philosophy could both promote evolution and in late life find that very science an insufficient basis for ethics. His agnosticism sought a middle path between nihilism and atheism. Like many of his generation (the young student H. G. Wells was one of his last pupils) Huxley disapproved of socialism, spiritualism, the unmarried relationship of George Eliot and G. H. Lewes; and he would not allow Oscar Wilde a return visit after an unasked-for call at his house.

Robert Jameson (1774-1854) Geologist and mineralogist. Jameson was a professor of natural history and keeper of the museum at Edinburgh University. He had a considerable reputation both at home and abroad but equally striking is the judgement on his lectures as 'a chaos of facts' (Thomas Carlyle) and (according to another visitor) Jameson himself as 'a baked mummy'. Darwin was terribly disappointed – indeed bored – by the lectures and deemed Jameson 'that old brown dry stick'.

Jean-Baptiste de Lamarck (1744-1829) The leading French naturalist of the late Enlightenment in France. The author of a great range of natural historical works, Lamarck became associated with a number of evolutionary ideas – such as the transmission of acquired characteristics – which for naturalists like Darwin were erroneous and grandiose. Lamarck's reputation suffered unfairly from then on.

W. A. F. Leighton (1805-89) Clergyman and botanist, most noted for his *The Flora of Shropshire* of 1841. Nora Barlow in her edition of the autobiography dates his death incorrectly as 1899.

Sir Charles Lyell (1797-1875) Scottish geologist and author of *Principles of Geology* (1830-33; 3 volumes). Lyell advocated a geology where natural forces currently in action were the same in nature and intensity to those active in the historical past. This has subsequently been described as 'uniformitarianism'. Darwin read his volumes while on the *Beagle* voyage and always praised them, including in the autobiography. Lyell's guarded response to an evolutionary explanation for the origins of human beings and his anxiety on the implications of an animal ancestry for man never diminished and became a great disappointment to Darwin.

Thomas Babington Macaulay, Lord Macaulay (1800-59) English historian and politician. Author of *The History of England* (Vols 1–2, 1849, Vols 3–4, 1855).

Thomas Robert Malthus (1766-1834) English clergyman and political economist. Author of *An Essay on the Principle of Population* (1798). Malthus's mathematical description of the perpetual mismatch between populations and their resources and the subsequent destructive actions of war, pestilence and famine was central to the structure of Darwin's argument in the *Origin of Species*.

St George J. Mivart (1827-1900) Self-taught natural historian and primate comparative anatomist. Mivart was a Catholic, a critic of Darwinism and someone for whom Darwin felt a considerable amount of personal enmity, not least when Mivart criticized the work of one of his sons.

Sir Roderick Murchison (1792-1871) A leading British geologist most noted for his construction of the Silurian System.

Sir Richard Owen (1804-92) Comparative anatomist. Owen worked first at the Royal College of Surgeons before becoming the overseer of the Natural History section of the British Museum and its establishment in South Kensington. Owen was a skilled proponent of the theory and practices of continental – especially French – comparative anatomy. Some of his schemes and strategies for undermining or dismissing Darwin's evolutionism led to a great deal of personal dislike.

William Paley (1743-1805) English natural theologian. A Fellow of Darwin's own college, Christ's College, Cambridge. Paley's works included *The Principles of Moral and Political Philosophy* (1785) and *A View to the Evidences of Christianity* (1794). Darwin read these as set texts for his BA examination and like almost all readers of Paley discovered their logic and their charm. Paley's natural theological system was, however, the first casualty of Darwin's own work.

Adam Sedgwick (1785-1873) Clergyman, geologist and Woodwardian Professor of Geology at Cambridge from 1818 until his death. A Yorkshireman,

Sedgwick represented a rugged combination of muscular Christianity and geological fieldwork.

Sydney Smith (1771-1845) English clergyman, writer and wit. One of the foremost English preachers of his day. As one of the founders of the *Edinburgh Review* (1802), his brilliant contributions were a major factor in the periodical's success.

Herbert Spencer (1820-1903) Philosopher, psychologist and sociologist. Very widely read in his day, not least in the United States, and now largely forgotten, Spencer wrote prolifically on the conjunction between environmental – almost Lamarckian – evolutionism and the development of complex social organization. Darwin in the autobiography obviously found him insufferable.

Christian Konrad Sprengel (1750-1816) German botanist. He stressed the role of insects and the wind in the cross-pollination of plants, but his observations were neglected until Darwin brought them to wider scientific attention.

Sir Edward Burnett Tylor (1832-1917) English anthropologist. Tylor's definition of culture in his book *Primitive Culture* (1871), and his further thoughts on animism and on the survival of past cultural practices, safely housed in more advanced societies, were highly influential. His *Anthropology* (1881) was the first textbook on the subject and he was also the first person to hold an academic position in anthropology when he became a lecturer at Oxford University in 1884. He was then Professor of Anthropology at Oxford from 1896 to 1909.

Alfred Russel Wallace (1823-1913) Naturalist. Wallace travelled in the Amazon between 1848 and 1852 and in the Malay Archipelago between 1854 and 1862. On the latter travels, he came close – but completely independently – to devising his own theory of natural selection. Darwin had enormous regard for Wallace's natural history, including his work on zoogeography and on mimicry in animal behaviour. This made Wallace's later commitments to spiritualism, socialism, anti-vaccination campaigns and the nationalization of land all the more strange in Darwin's eyes. At the very moment when evolutionary theory might start to look at human consciousness as developing out of animal forms, Wallace took the view that the existence and development of the mind could not be explained by natural selection. This defection into idealism (in the philosophical sense) was incomprehensible to Darwin. Both men, at least in public discourse, often described themselves as co-founders of the theory of evolution but their once shared road came to a great divide and never rejoined.

Josiah Wedgwood II (1769–1843) Darwin's uncle. Wedgwood's reassuring of Darwin's father that joining the *Beagle* expedition would be educative, respectable and morally safe had enormous repercussions in Darwin's life.

Gilbert White (1720–93) English naturalist and cleric. He was the author of *The Natural History and Antiquities of Selborne* (1789), a careful study of the seasons and animals, flora and fauna of his Hampshire parish.